CELL BIOLOGY RESEARCH PROGRESS

PROPERTIES LIVE AXOPLASM

CELL BIOLOGY RESEARCH PROGRESS

Additional books in this series can be found on Nova's website under the Series tab.

Additional e-books in this series can be found on Nova's website under the e-book tab.

CELL BIOLOGY RESEARCH PROGRESS

PROPERTIES LIVE AXOPLASM

O. S. SOTNIKOV

nova publishers
New York

Library of Congress Cataloging-in-Publication Data

Names: Sotnikov, O. S. (Oleg Semenovich)
Title: Properties live axoplasm / editor, O.S. Sotnikov (Lab. of Functional Morphology and Physiology of the Russia Academy of Sciences, Head of the Lab., Russia).
Description: Hauppauge, New York : Nova Science Publishers, 2015. | Series: Cell biology research progress | Includes bibliographical references and index.
Identifiers: LCCN 2015031084 | ISBN 9781634836845 (hardcover)
Subjects: LCSH: Axonal transport. | Neurons.
Classification: LCC QP363 .P766 2015 | DDC 612.8/1046--dc23 LC record available at http://lccn.loc.gov/2015031084

Published by Nova Science Publishers, Inc. † New York

CONTENTS

PREFACE

This book is devoted to the study of physiological properties of a living axoplasm, which, as it turns out, is rarely studied as a single organ. This study revealed a number of new physiological properties. The mechanism of simultaneous currents bidirectional axoplasm, disappearance streams and reappearance of the apical dendrites of pyramidal neurons in the brain during stress were revealed, as well as the causes of the similarity of nerve terminals regardless of their very different physiological functions. There exists proof of differences in presynaptic terminals of the postsynaptic. The reasons for the impossibility of treating transected axons using surgical sutures are discussed, and finally the cause of the absence of swelling properties in myelinated fibers with hypotension environment is solved. Some new approaches to the treatment of fibers and early manifestations of diastase nonspecific reactions are explored as well.

Keywords: axonal opposing currents axoplasm, submicroscopic fibers are invisible, and the discrepancy presinapsov postsinapsa, retraction bulb, structural dynamics of the living sensory endings property filamentous aggregation-tubular structures axoplasm dynamics notches Schmidt-Lanterman

INTRODUCTION

The purpose of fundamental science, in my opinion, consists not only in making new discoveries about the natural world and life but also to explain the potential mechanisms of the processes that we describe. What is particularly stimulating is scientific work in which the scientist manages to discover and provide proof of something unbelievable, that according to the opinion of the majority simply cannot be; in other words, turn prevailing opinions and laws on their head and confirm the existence of something hitherto considered science fiction. In this work we will examine the morpho-physiological, that most beguiling of processes. It is important to qualify that the particular interest is in the kinetic state of living structures and not their static state (construction, characteristics and so on).

In 1948 Paul Weiss and Helen Hiscoe discovered the movement of axoplasm. When the nerve was constricted in a certain area, axoplasm built up as though to prevent the further growth of the fiber (Weiss, 1959; 1967; 1970; 1972; Weiss, Taylor, Pilloi, 1962). This was a most interesting observation, the meaning of which has not been lost. However, the most ground-breaking discovery was made by Lilian Lubinska (1956; 1963; 1964). She voiced the idea and provided evidence that axoplasm can simultaneously flow in both distal and proximal directions. And regardless of how Weiss may object, or of what error Lubinska may have committed, the unbelievable turns out to be true. In this book, we describe for the first time how experiments carried out on living axons show the bi-directional movement of axoplasms with all their organelles; that is to say, something that cannot possibly be.

Experiments on "invisible fibers" demonstrate that apical dendrites disappear from compounds of the hippocampus under stress and appear again after two hours in their original place. This also looks like fiction—it would seem impossible. However, in experiments on living axons, we have turned fantasy into a genuine sub-microscopic reproducible process.

We have been able to demonstrate on living compounds and explain how and why all nerve endings, irrespective of their entirely separate functions, acquire the same structural form and the likely dissociation of pre- and post-synapses of living fiber, as J. Demoor (1896) and A.S Dogel (1896) attempted to prove. It is possible that this cannot be, but when living axons are shortened they give us cause to think otherwise. Pre- and post- synapses can detach, break off and form retraction bulbs.

The multiple forms of neuro-surgical methods of healing severed nerves of many clinicians lead to the idea that the liquidation of the diastasis of the nerve is impossible. But experiments on living organisms on the severed nerves of vertebrates and invertebrates with the application of blockers of nerve movement bear witness to the fact that inhibiting the retraction of axons is real, which means that stopping the dissociation of nerve stumps (diastasis) is also possible.

Another mystery of axons is the unbelievable phenomenon of the absence of swelling of myelinic fibers in hypotonic environments, something which the laws of physical chemistry disallow. Nevertheless, an analysis of the kinetics of living axoplasm, even here through the integral change of the dispersal of cytoskeletal proteins of axoplasm, enables the achievement of what seemed impossible.

In the work presented here, based on the irreversible retraction of axoplasm and its colloidal properties, we have tried to uncover some of those still existing secrets of living axoplasm. Drawing upon the huge volume of already known data about single cytoskeletal proteins, their interrelations and movement, we have tried to form an impression of the unity of exoplasm as an organ with its general unique characteristics that engender 'unsolvable' mysteries. They are impossible to explain on the basis of a conception of flowing liquid axoplasm. It is like a contractile myoplasm and is similarly governed by the situation, which we have come to view as the law of the simultaneous bi-directional retraction of axoplasm.

Chapter 1

THE SIMULTANEOUS FLOW OF AXOPLASM IN OPPOSITE DIRECTIONS

All organelles are in a state of constant motion. Many consider the main network of this engine to be neurofilaments (Nguyen, Shu, Sanada et al. 2004). To be more precise, the flow of axoplasm and the many proteins associated with microtubules at different levels of neurons are under investigation.

Many specialists are investigating the speed of axonal flows, including the mix of anterograde and retrograde flows together with viruses throughout all the nerves of the organism.

All this has led to the current understanding of axoplasm as liquid flows. However, it has been proven that a large role in creating the density of axoplasm is played by an abundance of neurofilaments. They constitute a significant mass of the skeletal organelles of axoplasm. It seems that neurofilaments make up the main network of the cytoskeleton.

All contemporary literature on molecular neurology is full of new, usually biochemical, information about the structure and movement of the independent and interrelated micro-molecules of domains and their interaction (Sotnikov, Vasyagina, Sergeeva, 2013). However, generalized information about axoplasm, about integral axon as a single organ possessing a range of particular physiological properties and functions, has not been sufficiently investigated (Nguyen, Shu, Sanada et al. 2004).

Neurofilaments determine the size of axons and the density of the spread of organelles. Along with microtubules they contribute to the achievement of the mechanical stability of the axoplasm (Huh, Laurer, Raghupathi et al. 2002) and consequently the speed of nerve passage. It is thought that the distance between single neurofilaments depends on the level of their phosphorylation and consequently the level of their aggregation (Li, Hu, Liu et al. 2009). However, it would seem that there are other physical properties, which enable the convergence of neurofilaments (Sotnikov, 2008). An excessive accumulation and change in their direction can provoke a range of unhealthy neurological forms. This is how many degenerative neurological illnesses start, amongst them Alzheimer's disease, which includes neurofilaments and the associated protein Tau, which is closely connected with microtubules (Mandelkow, Mandelkow, 2012; Zaichick, Bohannon, Hughes et al. 2013; Feller, Cronin, 2014). The main engine of neurofilaments is considered to be the myosin Va, associated with neurofilaments (Jung, Chylinski, Pimenta et al. 2004).

By analyzing the electron-microscopic structure of the axon, the important role of the microtubules of the cytoskeleton should be noted. They are the main agents of the mitochondria and other organelles involved in the anterograde movement of regeneration material and the formation of mechanical properties of the axoplasm (de Forges, Bouissou, Perez, 2012). Microtubules are considered to be an important factor in the organization of axons and their mechanical properties (Piao, Wang, Xu et al. 2004; Qiang, Yu, Liu et al. 2010). It is important to note that the length of microtubule does not change their number although they are closely related to each other (Rao, Garcia, Miyazaki et al. 2002). We know the importance of the property of self-organization of microtubules into bundles when the external environment, ionic strength, and specific agents are changed (Bird, Lieberman, 1976; Hamon, Savarin, Curmi et al. 2011). Dissociation of microtubules and neurofilaments can be triggered experimentally by colchicine or cytochalasin (Sotnikov, Vasjagina, Sergeeva, 2013).

All of these structures and macromolecules are in constant motion. After a long discussion it has now been shown that both short and long microtubules are involved in neuronal transport (Baas, Mozgova, 2012). The motor activity of single polymers of the cytoskeleton has been widely discussed. Depending on their speed, the organelles are divided mainly into three categories: fast, medium and slow (Baas, Buster, 2004; Baas, Nadar, Myers, 2006; Baas, Yu, 2012). A general pattern of stops and subsequent rapid movement of the neurofilaments has been disclosed, which is why for a long time they were considered to move slowly (Roy, 2014). The possibility of changing the direction of the anterograde flow of the axoplasm to retrograde has been described (Chowdhury, Coats, Neis et al. 2010; Mandelkow, Mandelkow, 2012), although simultaneous bidirectional motion has never been demonstrated. There has been a lot of experimental work on the reversible retraction of axoplasm (Salinas, Bilsland, Henaff et al. 2009; Kato, Kobayashi, Inoue et al. 2011; Kramer, Greco, Taylor et al. 2012).

All of these studies were inclined towards notions of liquid flows in axoplasm and towards its various velocities. However, "liquid" axoplasm cannot maintain a cylindrical shape without the endoneurium and the cell membrane, especially since neurolemma is a fluid formation. All contemporary literature of molecular biological neurology is full of information, usually biochemical, on the structure and motion of independent and interconnected macromolecules, domains and their interaction (Sotnikov, Kokurina, 2013; Sotnikov, Vasjagina, Sergeeva, 2013). At the same time, generalized data specifically on axoplasm as a single body, possessing a range of its own non-electrical properties and functions, are poorly understood.

Clearly, the factors forming the cytoskeleton of axoplasm are polymers and their many forms of associated proteins, providing the stable structure of the axons. The best known among the associated proteins are the motor proteins: kinesin, dynein and also tubulin, which are associated proteins (Wang, Chen, Chen, 2012; Velazquez-Bernardino, Carcia-Sierra, Hernandez-Hernandez et al. 2012). However, the decisive role of associated proteins is not usually emphasized and the role of a significant number of various associated proteins has not been fully explored.

METHOD

Although many single polymers have been studied, the total non-electric function axoplasm as a whole remains unknown. The flow of axoplasm is often compared with blood, although blood vessels are known to have a more mechanically tight endothelium, reticular connective tissue fibers of the pericapillary cells and other reinforcing structures. Apparently, the neurolemma and surrounding connective fibers of the endoneurium, the vessels along which fibers are distributed, and the neighboring tissue imparts the geometric, mechanical shape of the fibers. Nevertheless the shape of neurites change radically during ontogeny.

It is of interest, how the axoplasm would behave when devoid of all these factors stabilizing it. After all, it would seem that axoplasm currents can flow only in certain cases. However, after the neural cells isolated from ganglia were completely deprived of the connective tissue and surrounding structures, they still substantially retained their shape (Figure 1). It follows therefore that they must have been surrounded only by axolemma. However, as is well known, the monolayer membrane is a liquid substance. This implies that the shape of the fiber is retained mainly due to the relatively elastic properties of the axoplasm. Consequently, axoplasm flow cannot be compared with blood flow.

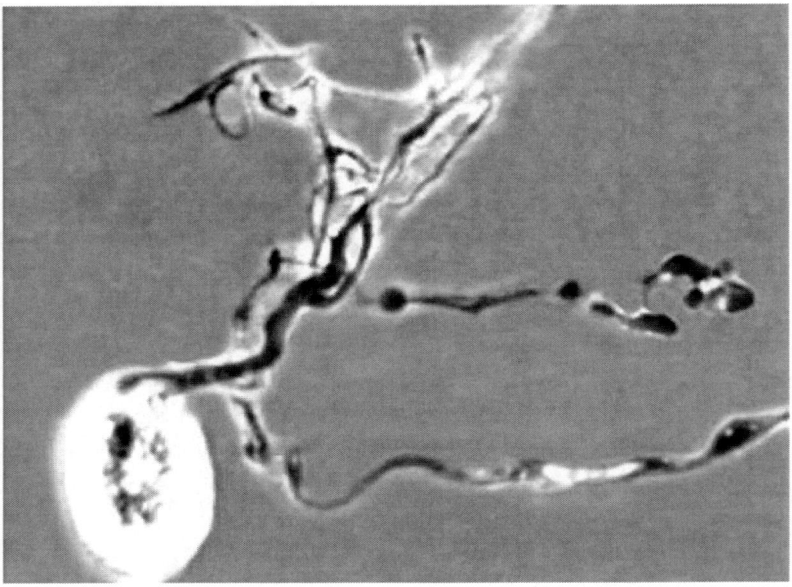

Figure 1. Single isolated ganglion neurons of the mollusc Lymnaea stagnalis, devoid of connective tissue. Intravital microscopy. Phase contrast. Vol. 40 Ph, approx. 10.

The model of dictyostelia, a construction of multicellular organisms has been studied in great detail, but the physical properties of axoplasm, its cytoskeleton, retaining the shape of exposed appendages, still needs further development (Zenbush, 1982). It remains unclear how the single molecular motor cytoskeletal structures in the overall transport of axoplasm operate together, what the mechanism of the incredible, simultaneous bi-directional flow of axoplasm is, and, finally, what is the total motor mobility of neuroplasm (Baas, Mozgova, 2012; Sotnikov, Vasyagina, Sergeeva, 2013; Sotnikov, Vasyagina, Sergeeva, 2014 a, b).

Taking into account an analysis of the literature, it is necessary to note the incredible combination of the many functions of neuroplasm: its fluidity, motor activity and mechanical strength. Such characteristics are present, possibly, only in living systems such as myoplasm and axoplasm.

Exploring the transport of integral axoplasm is only possible on living single axons of isolated unmyelinated fibers, deprived of endoneurium. Research was carried out for 2-8 hours. In general, living fibers are capable of conducting electrical impulses for 24-48 hours (Gaudet, Popovich, Ramer, 2011). Inverse phase and interference microscopes were used. It should be noted that even the most modern immuno-histochemical studies have to apply roughly coagulating clamps, which often cause long-known artifacts and are forced to use them as new structures that are of great functional importance (Perrin-Tricaud, Rutishauser, Tricaud, 2007; Buttermore Dupree, Cheng et al. 2011; Susuki, Raphael, Ogawa et al. 2011; Terada, Saitah, Ohno et al. 2012).

It should be noted that in the most successful preparation you can get neurons with one or two appendages and a considerable degree of ferning, which exceeds the degree of branching of appendages from the preparations of neurons obtained by the key scientists of neurohistology (Deiters, 1865; Deiters, Guillery, 2013) carried out, it must be said, on fixed preparations. The speed of retraction of axons and the volume of change of the soma of axons was determined.

Also monitoring was carried out by means of transmission electron microscopy. The body of neurons were undamaged even during centrifugation (Figure 2). Single parallel microtubules were studied in detail. Axons contained up to 6 connected microtubules (Figure 3).

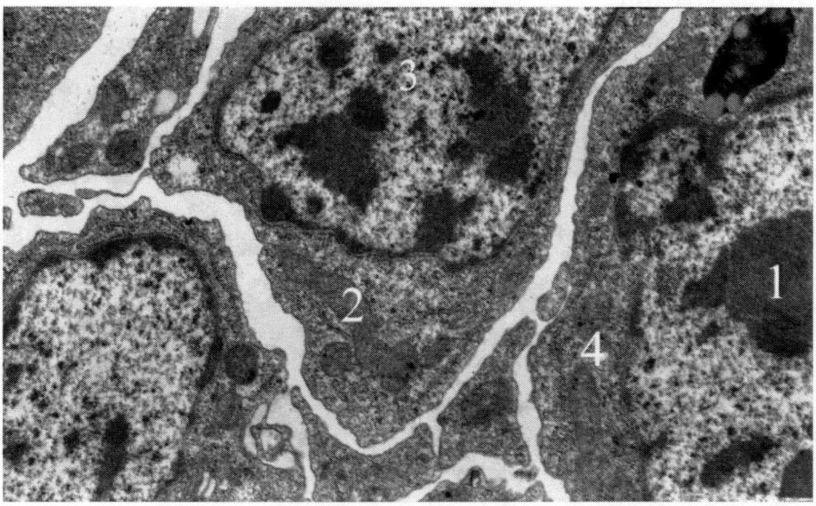

Figure 2. Neurons of undamaged ganglion of the mollusc Lymnaea stagnalis without pronase treatment. 1 - nucleolus; 2 - mitochondria; 3 - the core; 4 - neuroplasm. Electron microscopy. Mag. 40,000.

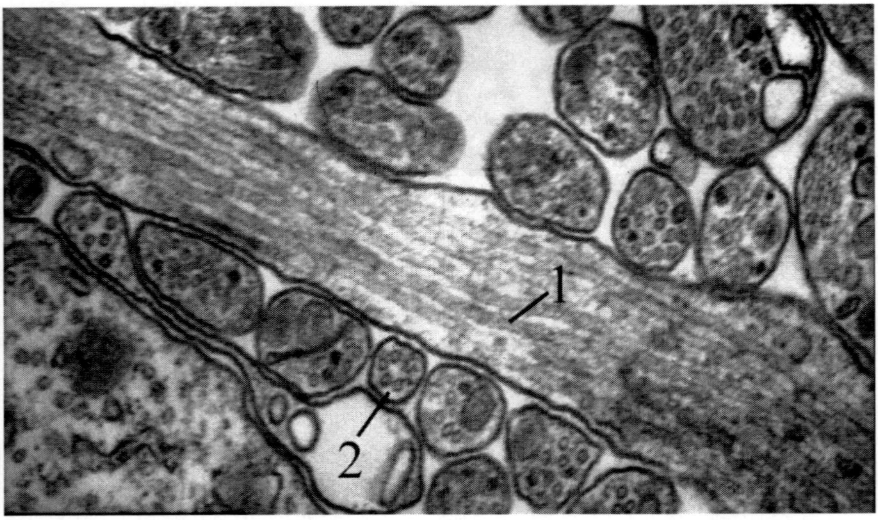

Figure 3. Microtubules of thin nerve fibers from the dissociated ganglion of the greater pond snail. 1 - microtubule; 2 - microscopic glial-free fiber. Electron microscopy. Mag. 70,000.

Typically flat parallel microtubules have protrusions on their surface of different mass in the form of hillocks, clavate or bifurcated projections (Figure 4) that resemble “handles,” and apparently are microtubules, that is, associative proteins.

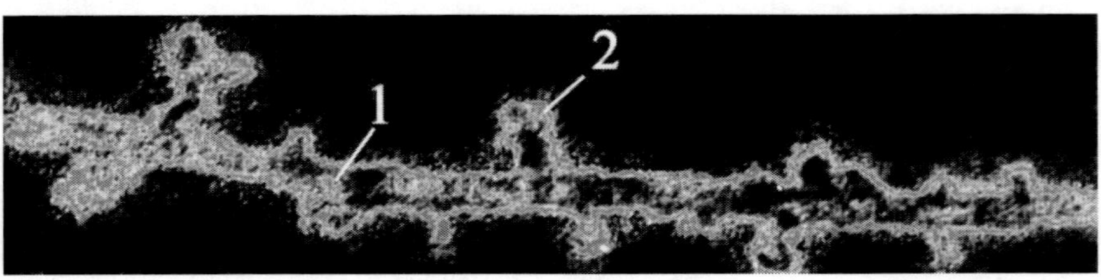

Figure 4. A microtubule of axoplasm with its associated proteins. 1 – microtubule; 2 – associated proteins. Electronic microscopy. Mag. 70 000.

Transport of the organelles of axoplasm is usually attributed (Figure 5) to these proteins. Many authors suggest that organoids are associated with microtubules and move independently along them as on rails using mobile associative proteins.

The incline of the associated peptides can sometimes even determine the direction of motion of the moving organelles. However, the idea of the mobility of microtubules themselves, their relationship with the filamentous cytoskeletal apparatus and the ability to organize fascicles is contradictory and not sufficiently clear.

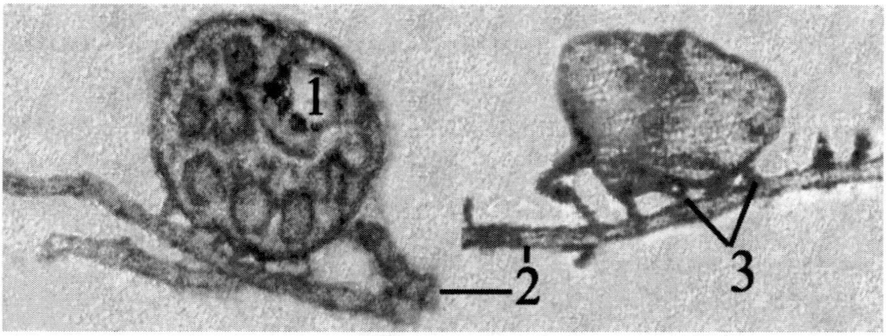

Figure 5. Single axoplasmatic organelles, supposedly moving along microtubules. 1 - multivesicular bodies; 2 - microtubules; 3 - tubular-associated proteins. Electron microscopy. Mag. 70,000.

Adjacent microtubules often use associated proteins as stabilizing transverse bridges, so that, together with microtubules, associated proteins form a densely organized net-like structure (Figure 6).

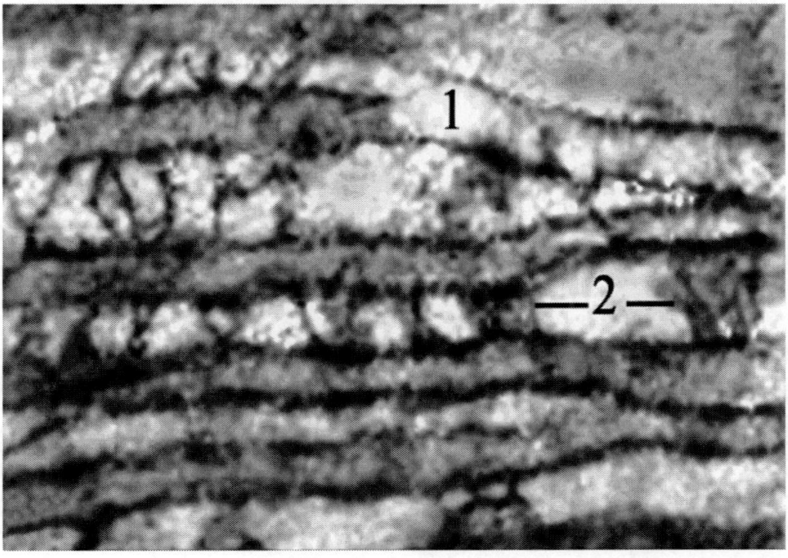

Figure 6. Cytoskeletal associated proteins, combining microtubules in a structural group. 1 - microtubule; 2 - microtubule associated proteins (bridges). Electron microscopy. Mag. 70,000.

Such bonds between microtubules can be broad with a high optical density. Settling across, it is as if tubular polymers bond into a single filamentous-tubular cytoskeletal system. Between the microtubules and neurofilaments a picture is also created that forms a kind of relative mechanical stability. If microtubule associated proteins are often characterized by their massiveness, blur and variation in direction, then the associated proteins of neurofilaments are thinner and arranged exactly perpendicular to the neurofilaments. Given this, they clearly resemble a ladder construction (Figure 7).

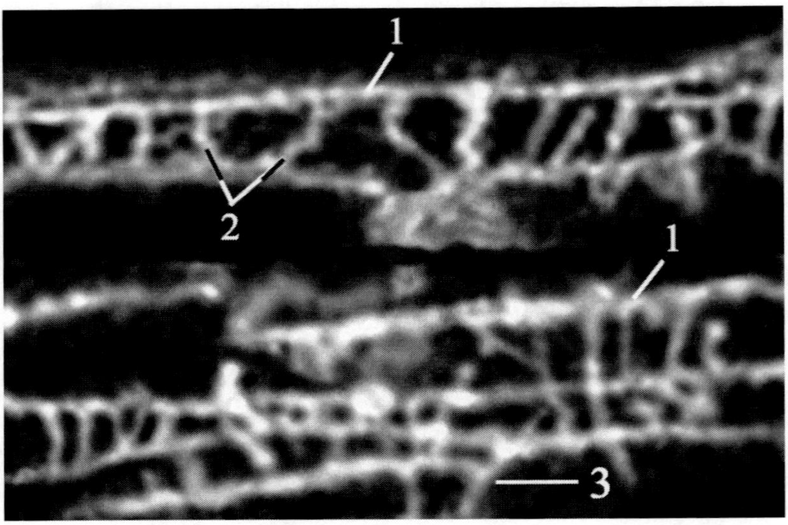

Figure 7. Mechanical fixing of neurofilaments bound in paired fascicles. 1 - neurofilaments; 2 - neurofilament-associated proteins; 3 - filamentous cytoskeletal network. Electron microscopy. Mag. 70,000.

There are all forms of cross-bridges not only between independent microtubules and neurofilaments but also simultaneously between these organelles and neurolemma. The formation of biological structures of axoplasm usually occurs with the increasing concentration of the dispersed phase from true sols to solid systems, but the law of any creation requires that unrelated conglomerate structures are bonded and adhered before they are converted into structure (Figure 8). Only the presence of the adhesive material allows the creation of “art.”

Figure 8. Groups of unorganized building blocks (a), with the application of adhesion (b) are transformed into an organized structure. 1 - adhesive - cement weight; 2 - organized structure.

That said, the cisterns of the endoplasmic reticulum have an original appearance. They consist of wide tubules and separate cisterns. On their surface "handles" can also be traced (Figure 9).

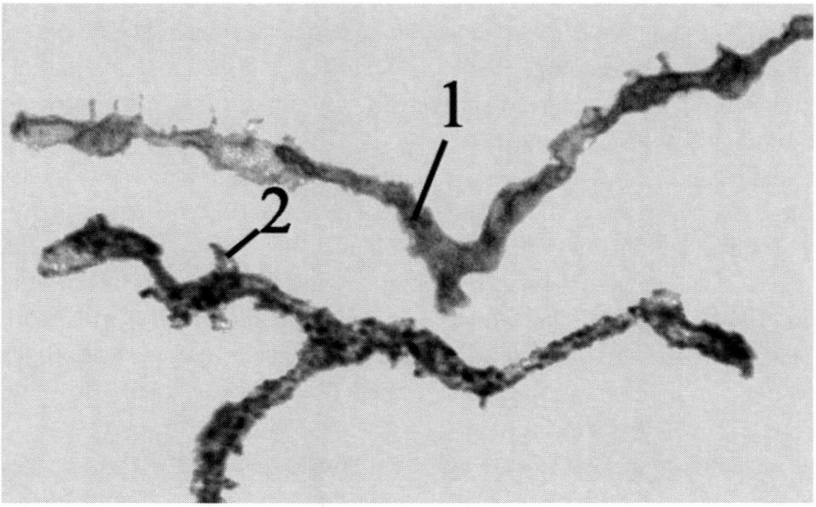

Figure 9. Endoplasmic reticulum of neuroplasm. 1 - endoplasmic tube; 2 - associated proteins of endoplasm. Electron microscopy. Mag. 70,000.

Electron microscopic analysis requires a description of another general cell organelle - the endoplasmic reticulum. It consists of wide individual tubes and tanks. Contact of endoplasmic reticulum cisterns with elements of axoplasm has already been noted repeatedly (Masurovsky, Peterson, Crain et al. 1981).

The possibility of endoplasmic reticulum associated proteins is also shown. So far, however, mechanical structural reorganization of endoplasmic reticulum in the course of its life remains poorly studied. It may contain ribosomes (granules) on its surface (Taylor, Grim, Stout, 2007). However, the ratio of smooth and rough endoplasmic reticulum with other organelles is within norms even under external action (Lewin, Kassimeris, Lingappa, 2011).

The endoplasmic reticulum is a dynamic system. The relationship between it and the ultrastructures has been investigated (Senda, Fujita, 1987). The endoplasmic reticulum can be used by transport vesicles (Karcer, 1995). The synthetically important role of the endoplasmic reticulum in the metabolism of lipids and fatty acid wax is known (Becker, Deamer, 1996). Although the endoplasmic reticulum envelops neurofilament and microtubule fascicles and permeates their network, all the layers of the cytoskeleton, the mechanical role of the organelles as a single integral body remains unclear (Sotnikov, Vasyagina, Sergeeva, 2013).

The tubes of the endoplasmic reticulum and cisterns cover, as it were, the flat parallel filamentous-tubular elongated three-fold woven, often anastomosing, networks. These morphological descriptions related to axons as a whole, in our opinion, have a significant functional physiological significance. The mechanical properties of tubules of the endoplasmic reticulum have not been explored.

Against the background of filamentous, tubular nets and adhesion proteins, axoplasm nerve fibers look like a dense netlike formation, likely to be a mechanical obstruction for liquid, and especially for the visco-elastic substance of axoplasm (Brown, Henning, Wellman, 2005). No wonder that P. Weiss (1948) considered the axon as a solid column that spreads in the form of axoplasm in a growing neuron. In the discussion about the physical properties of axoplasm our observation of the severed fiber not only ceases to flow but also retracts away from the cut.

The fact is that nowadays there are hundreds of published works relating to axonal flow and transport mass of single proteins with different functions. Apart from a few types of axonal flows that are fast, slow and ones that demonstrate interrupted flow, they are capable of interchangeable anterograde or retrograde transportation of axoplasm. However, large (up to 25 nm in diameter) and long microtubules and neurofilaments undoubtedly possess mechanical elasticity and resist the transport of axoplasm. In connection with all the structural complexity of intertwining organelles, the question arises: can axoplasm filled with netlike cytoskeletal protein polymers flow (overflow) through a natural three-dimensional mesh (Figure 10), or more specifically, flow simultaneously in opposite directions?

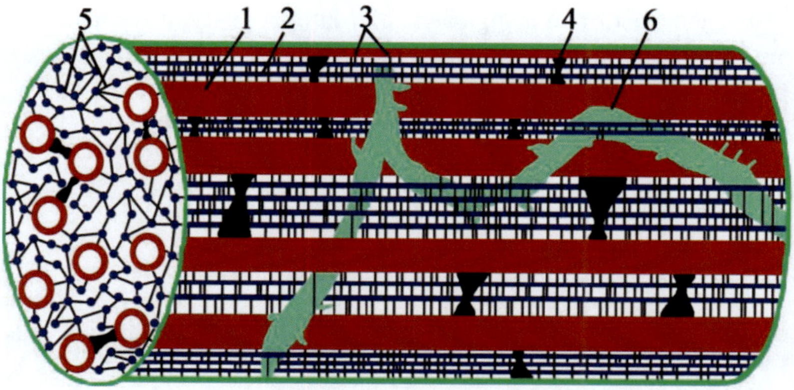

Figure 10. A diagram of the mechanically stable axon, devoid of membranes and being a single axoplasm forming filamentous, tubular structures fastened by their associated protein adhesion bridges (Diagram). 1 - neurotubules; 2 - neurofilaments; 3 - proteins associated with the skeleton; 4 - proteins associated with the axolemma; 5 - proteins associated with neurotubules and neurofilaments; 6 - endoplasmic reticulum.

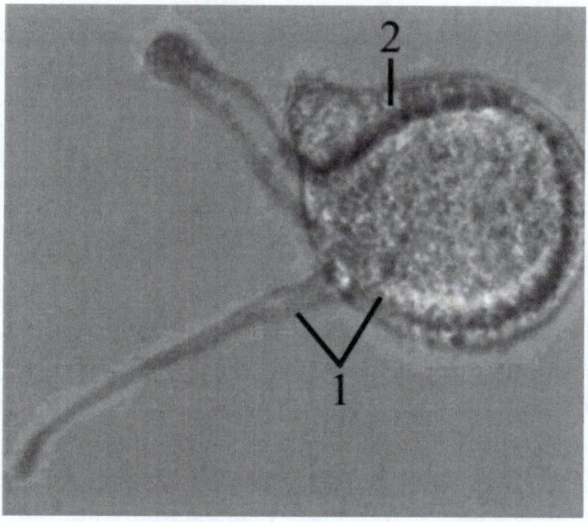

Figure 11. An artificial blending of images of the neuron before (1) and after (2) retraction of the appendages.

According to our elementary observations with living nerve transection, no leakage of axoplasm from amputated nerve fibers has been observed. On the contrary, axoplasm bi-directionally invaginates slowly in opposite directions from both cuttings simultaneously (Figure 11). The proximal segment of the fiber often invaginates fully into the body of the neuron, while its volume increases slightly. In the process of shortening the appendage, an increase in volume of the cell body to 57% was observed (Figure 12).

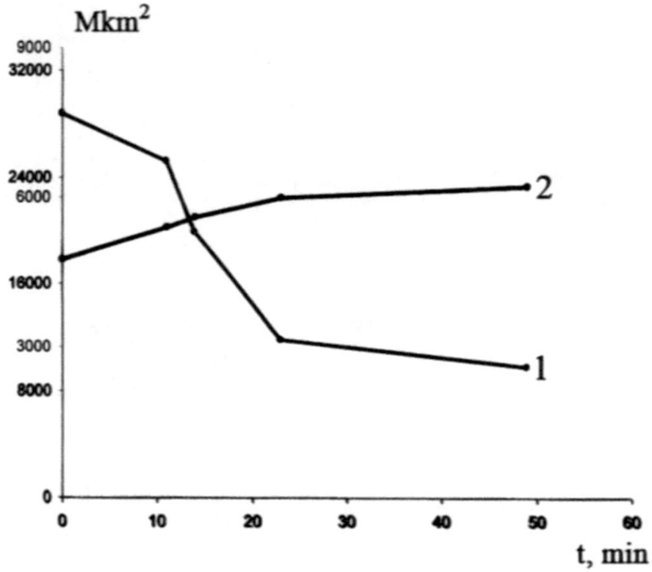

Figure 12. The proportional decrease in area of the optical severance of the appendages (1) and increase in the area of the severed soma neuron (2) (over time).

Since axoplasm does not flow into the area of the cut and becomes reduced, most likely, it cannot be called a liquid substance, although the propensity to thoroughly mix is only weakly connected to water, like any protein in unusual circumstances, it possesses it. (Kaivarainen 1980; Frolov, 1982; Friedrichsberg, 1984).

The ideas described do not exclude the vast experience of molecular biology studies on axonal flows and individual macromolecules, but may allow a look at the problem of axoplasm from another, unfamiliar angle.

Upon living axonal transection two points are found. Since axons invaginate, it would appear that they are under some degree of tension before the transection, and secondly, that the direction of movement of the axoplasm stumps is initially in the opposite direction, that is, it has a mechanical tone, contrary to the tension of the retraction. In the living axons, there is no doubt that the axoplasm of both stumps after transection of the fiber is conversely drawn into each stump of the severed nerve. This said, neurolemma behaves the same way as a liquid formation, becoming a part of the axoplasm.

These simple observations do not pre-empt any change in theory. However, we have traced patterns in the direction of transport of axoplasm depending on points of support (adhesion, cohesion) of the preparation. It would seem that such basic common knowledge was obtained as early as 1948 (Weiss, Hiscoe) with a simple method of the morphological overlay of ganglions on to the nerve fibers. This led to the accumulation of axoplasm at its ends, which indicated the presence of an anterograde flow in the axon (down) of axoplasm from the synthesizing center (nerve cell) to the periphery. On this subject, a lot of work was carried out (Weiss, 1959; Weiss, 1970).

It has been exactly 50 years since the publication of the famous article (report) Lilian Lubinska (1964) "Axoplasmic Streaming in Regenerating and in Normal Nerve Fibres." It was recognized by the international scientific community to be the result of a number of pre-publications of the authors (Lubinska, Niemierko, Oberfeld, 1961; 1963), which showed ample evidence of the simultaneous bi-directional flow of axoplasm. J.C. Eccles (1964) strongly supported the seemingly unusual study of L. Lubinska.

The mechanism of the simultaneous "flow of neuroplasm" of the same fiber in opposite directions has been discussed in many reviews (Becker, Deamer, 1996). The morphological method of the study of retrograde flows using peroxidase of cochlearia armoracia has become traditional (Kristenson Olsson, 1971), based on the cellulopetal transport of neuroplasm, and other studies (Skibo, 1980; Salinas, Bilsland, Henaff et al. 2009; Kramer, Greco, Taylor, et al, 2012). The bi-directional flow of axonal organelles and molecules in nerve fibers has been repeatedly proven with the help of radio-authography (Lazek, 1961). However, the mechanism of this phenomenon and the simultaneous regulation of the direction of axoplasm flows remain unexplained. The dual anterograde and retrograde movement of viruses in the axons is now widely studied by virologists in connection with the development of gene therapy of degenerative nerve diseases (Kato, Kobayashi, Inoue et el., 2011; Kramer, Greco, Taylor et al. 2012; David, Saied, Charles et al. 2012; Zaichick, Bohannon, Hughes et al. 2013). There is scientific evidence that actin and dynein can carry traffic in both an anterograde and retrograde direction, although the mechanism of this phenomenon remains unclear (Alves, Burzynski, Delalande et al. 2010; Liu, Nadar, Kozielski et al. 2010;

Falnikar, Tole, Baas, 2011). A large number of scientific publications are devoted to the reversible retraction of dendritic spines under titanic stimulation and other experimental effects (Cook, Wellman, 2004; El Maarouf, Kolesnikov, Pasternak et al. 2005; Izquierdo, Wellman, Holmes, 2006; Hill, Hillard, McEwen, 2011; Eiland, Ramroop, Hill, et al. 2012). Work has appeared that convincingly demonstrates reversible reduction in the length of the apical dendrites and arborization in pyramidal neurons of the cortex, hippocampus and other parts of the brain during hibernation (Popov, Bocharova, 1992; Popov, Bocharova, Bragin, 1992), the action of anesthetics (Al-Jahdari, Saito, Nakano et al. 2006) repeated stress (Billuart, Winter, Maresh et al. 2001; Cerqueira, Mailliet, Almeida et al. 2007; Radley, Rocher, Rodriguez et al. 2008) and a number of pathological effects (Martínez- Tellez, Gómez-Villalobos, Flores, 2005; Miller, Morrison, McEwen, 2012).

Observation of axoplasm flow is made easy thanks to the study of the reduction of axons deprived of their membrane. Naturally, the reduction of entire neural appendages simultaneously demonstrates retrograde molecular transport of the mass of neuroplasm. Neuroplasm moves as part of the growing or retracting appendage as a single body (Jung, Chylinski, Pimenta et al. 2004). Despite the large amount of convincing data on the simultaneous bi-directional flow of axoplasm, a hypothesis of the mechanism of this unique phenomenon remains unproven or unfounded.

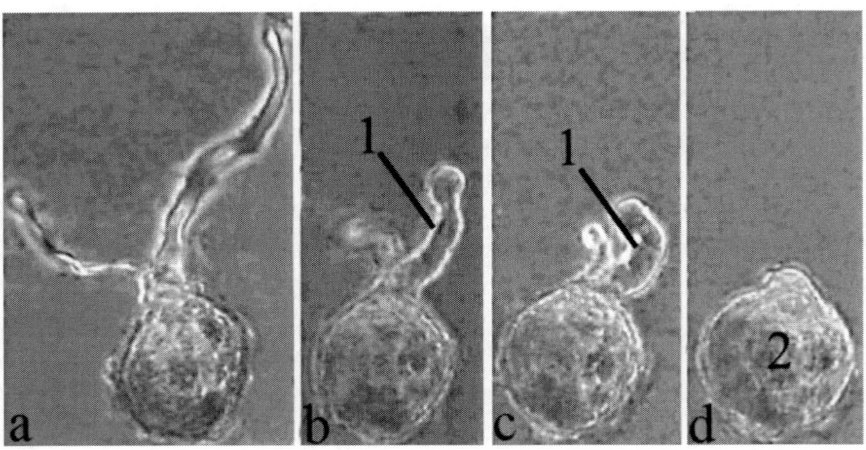

Figure 13. The retraction of nerve appendages of a denuded neurite into the cell body. Figures a-d show dynamics of the process; 1 - retracting appendages; 2 - body cells with appendages, drawn into the body of the cell. Intravital microscopy. Phase contrast. Continuous video. Vol. 40 Ph, approx. 10.

The method of forming "influxes" of axoplasm to zones of pressured or severed nerves, which served as the primary evidence of the anterograde transport and bi-directional flow of axoplasm, remains a mystery (Lubinska, 1964).

The simultaneous bi-directional reduction of axoplasm was started by us on "naked" axons with many appendages. It was found that the appendages of these axons always move toward the cell body. It became clear that the mass of the soma of a neuron and its adhesion dominate the appendages. During the process the appendage is usually reduced in size and moves in the direction of the body, invaginating into it (Figure 13).

Most appendages of the preparation are fully invaginated into the body of the cell. Part of the cells typically lose all secondary appendages if they do not exceed 30 mcm (Figure 14).

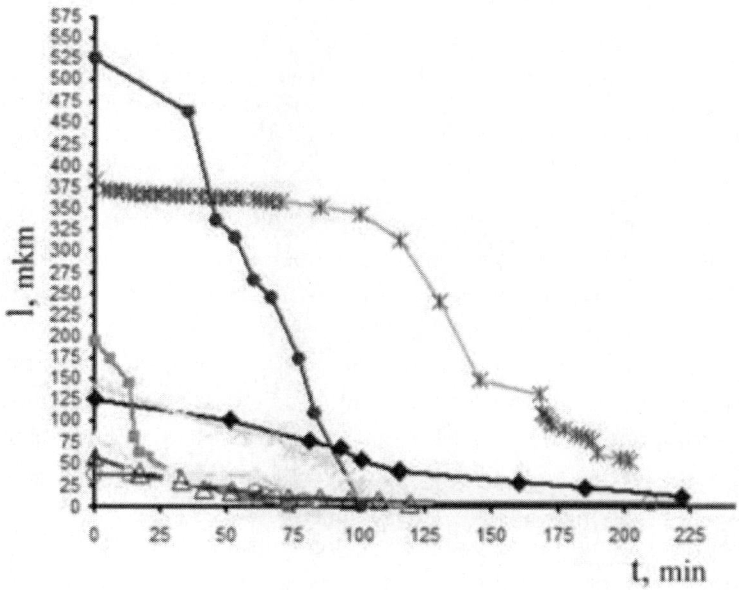

Figure 14. Graph of the speed of reduction of traumatized nerve appendages of various length.

If the point of adhesion is changed, not fixed on the cell body but at the end of the appendage, the direction of motion of the preparation is sharply reversed (Figure 15). However the axoplasm still moves to the body of the neuron.

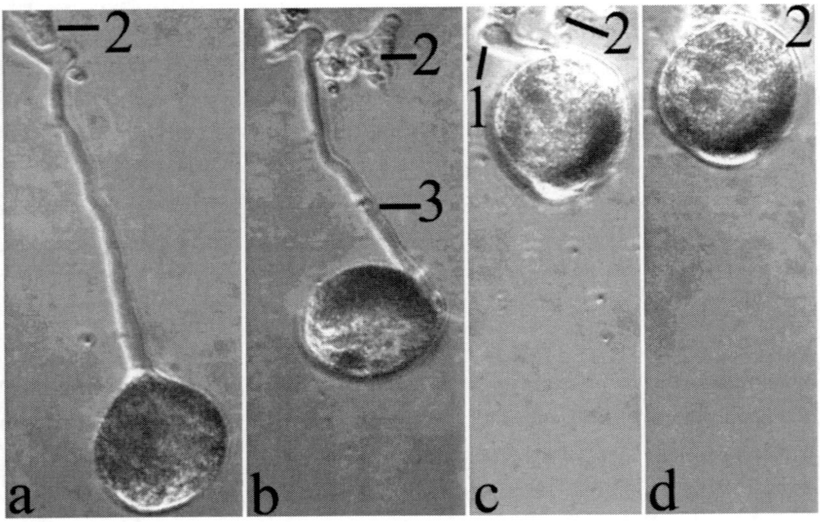

Figure 15. The movement of the neuron to the opposite, cellulofugal side when the point of adhesion is transferred to the fiber end, the mass of the axoplasm moves to the soma of the neuron. a-d - stages of the process; 1 - retraction of the bulb; 2 - Additional cells, providing a new point of adhesion of the drug; 3 – the shrinking axon. Intravital microscopy. Phase contrast. Continuous video. Vol. 40 Ph, approx. 10.

As has been shown in the experiments of Peter Frantsevich Lesgafta (1968) possessing a retraction tone, muscle fiber always shifts towards support (the point of adhesion). So we decided to test the effect of changing the position of the support point on the direction of movement of axoplasm. By moving the point of maximum adhesion, we got the opposite direction of movement of axoplasm. Thus, if the point of adhesion is placed in the middle of the preparation (Figure 15), the axoplasm will shift toward the center in two directions, toward the point of adhesion. This increases not only the size of the cell body, as usual, but the volume of the appendage ends.

Myologists know that if an injury occurs across the muscles, then at the site of the injury, just like the zone of adhesion of an axon, the muscle tissue swells, that is, the muscle mass strengthens in the area of the injury. It is reminiscent of the counter movement of axoplasm to the area of adhesion.

Therefore, there is another simultaneous bi-directional movement of axoplasm. During this process the point corresponding to adhesion becomes thinner, which indicates the partial direction of movement of the axoplasm in opposite directions (both to the end of the preparation and to the body of the neuron).

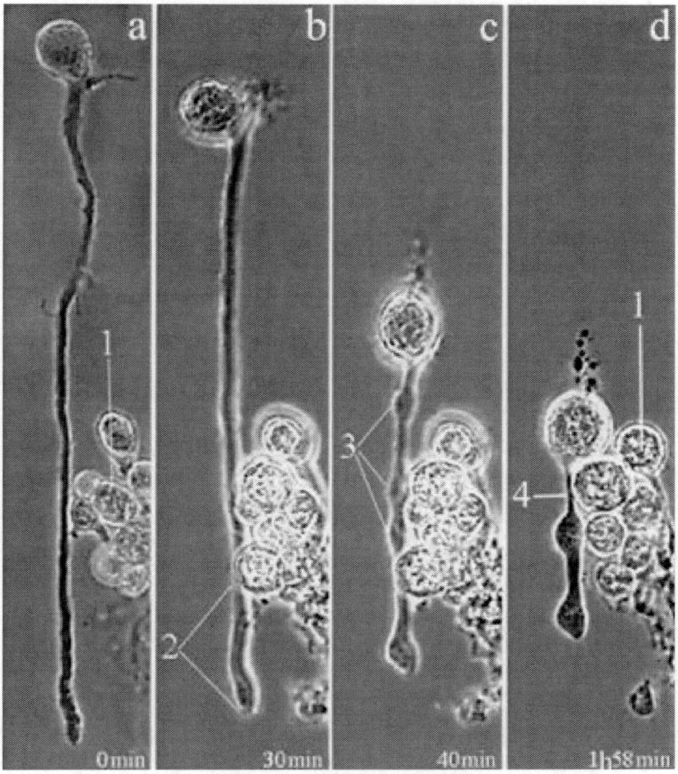

Figure 16. The movement of the mass of axoplasm in two opposite directions (to the soma and the end of the fiber). The movement of the ends of the preparation in the opposite direction. a-d - stages of the process; 1 - fiber adhesion area, formed by the additional cells; 2 - slow speed of the reduction of axoplasm at the peripheral end of the preparation; 3 - varicose deformation of the axon; 4 - thinning axoplasm mass in the middle of the preparation in the area of adhesion. Intravital microscopy. Phase contrast. Continuous video. Vol. 40 Ph, approx. 10.

A second series of experiments consisted in completely liquidating the point of adhesion. At the same time we isolated a fragment of floating appendage that had no adhesion points (Figure 16). It turned out that the mass axoplasm increased, moving from both ends toward the center, towards each part of itself (Figure 17), which is bi-directionally.

At this point the central section thinned, as in the previous experiment, indicating a partial offset of the axoplasm in opposite directions from the center (Figure 17). Movement of the axoplasm in opposite directions leads to a thinning of the appendage.

In the third series of experiments, the point of support of the elastic retracting fiber was fixed from both sides (Figure 18). It was supposed to eliminate the movement of the preparation, that is, a change in length of the nerve cell appendage. However, we were wrong.

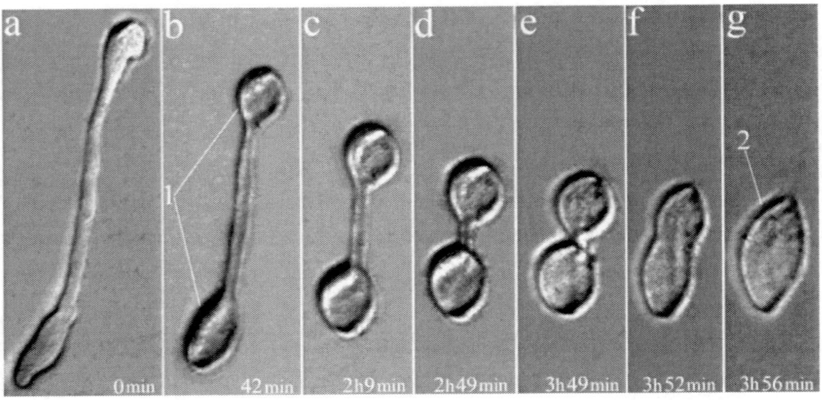

Figure 17. Bi-directional retraction of the lateral mass of the axoplasm of isolated nerve fibers towards each other in the absence of adhesion points at the ends of the preparation. Thinning of the central section of the appendage, indicating a partial move of the axoplasm in opposite directions from the center. a - g - stages of the process; 1 - retraction bulbs; 2 - thinned portion of the fragment; 3 - axoplasm merging into a single mass of spherical shape. Time - from start of observation. Intravital microscopy. Phase contrast. Continuous video. Vol. 40 Ph, approx. 10.

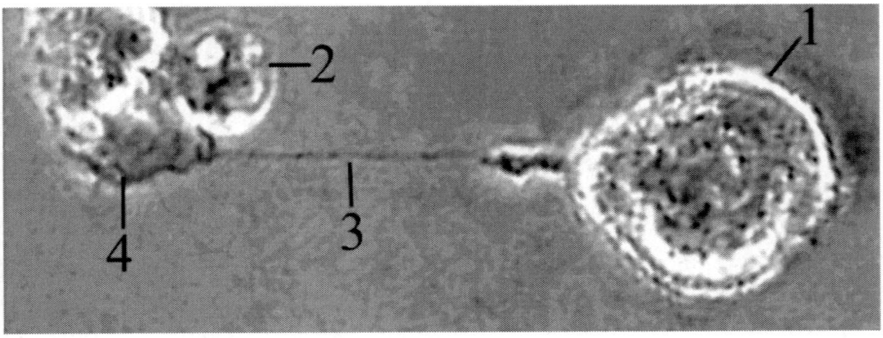

Figure 18. Dynamics of the isometric (bulk) retraction of the preparation. Thinning of retracting nerve appendage in the presence of adhesion points at both ends. 1 - an increase of the cell body and the opposite end of the fiber (simultaneous neuroplasmic flow in the opposite direction); 2 - the remnants of adjacent cell medium providing a second point of adhesion of the preparation; 3 - a sharp natural thinning of the elastic appendage, losing the limits of optical visibility; 4 - increase in the retraction bulbs. Intravital video microscopy. Phase contrast. Vol. 40 Ph, approx. 10.

Securing the ends of the preparation did not change its length. But they thinned dramatically. At the same time there was an increase in the periphery (volume) of the axoplasm and the cell body to the ends. That is, we are again talking about the translocation of axoplasm mass in opposite directions.

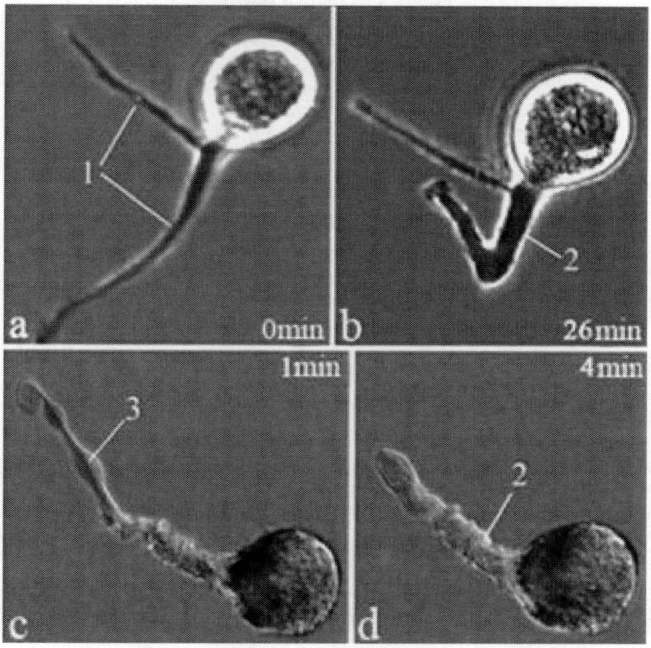

Figure 19. Expansion of the axon upon delayed retraction. a-b - a thickening of one of the appendages of the neuron; c-d - thickening of the appendage during its retraction; 1 - appendages of the neuron; 2 - their thickening; 3 - developing varicosity of the axon. Intravital video microscopy. Phase contrast. Vol. 40 Ph, approx. 10.

This solves the debate on an important issue for neurohistology on plasma. This phenomenon we have called isometric or volume retraction, i.e., axoplasm retraction occurs not as a result of reducing the length but by a sharp reduction in its diameter.

This new phenomenon, in our opinion, has real significance for the non-electrical functions of the axon. The nerve appendage may expand, increasing the diameter. In this case it slows down its movement to the cell (Figure 19). In other words, axoplasm of the denuded axon can move in all directions (Figure 20). See isometric contraction of axons in tissue culture (Figure 21).

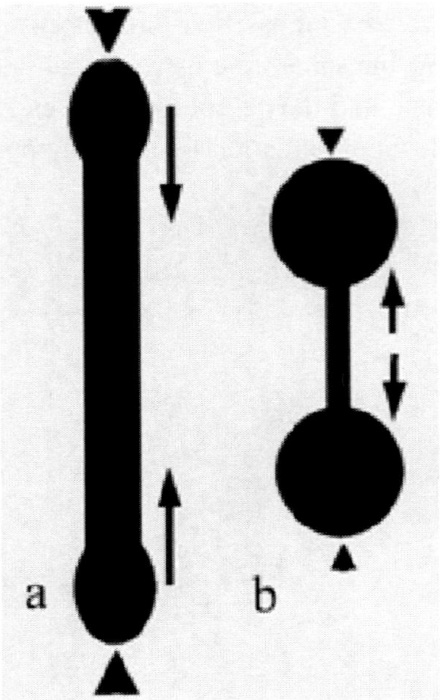

Figure 20. Schematic diagram of the overall concurrent axoplasm flows in opposite directions. a,b - the dynamics of the process. The arrows indicate the direction of the axoplasm flow; ▲ - triangles signify the maximum and the relative degree of axon contraction.

A prerequisite of this type of retraction is to secure both ends of the preparation. But it should be noted that in the nervous system all the outgrowths of neurons have two fastening points, and thus are capable of thinning, that is, to undergo isometric contraction. In other words, axoplasm has the ability to change its volume due to a change in length, moreover, not only by narrowing the axon, but by means of its extension. Thus, it is possible to uncover the secrets of the simultaneous bi-directional retraction of axoplasm. Although we found no axoplasm flows, it became possible to show the muscle-like simultaneous movement of axoplasm mass in opposite directions. We similarly demonstrated the possibility of not just constriction but also expansion of the axon. Through the use of a preparation of living axoplasm, enclosed only in neurolemma and a cytoskeletal network, it was possible to demonstrate "flows" of axoplasm and their retrograde tension and muscle-like contractility. It has been shown that the direction of "flows" of axoplasm depends on the zone of support and the points of adhesion of the preparation, and always has simultaneous bi-directional movement.

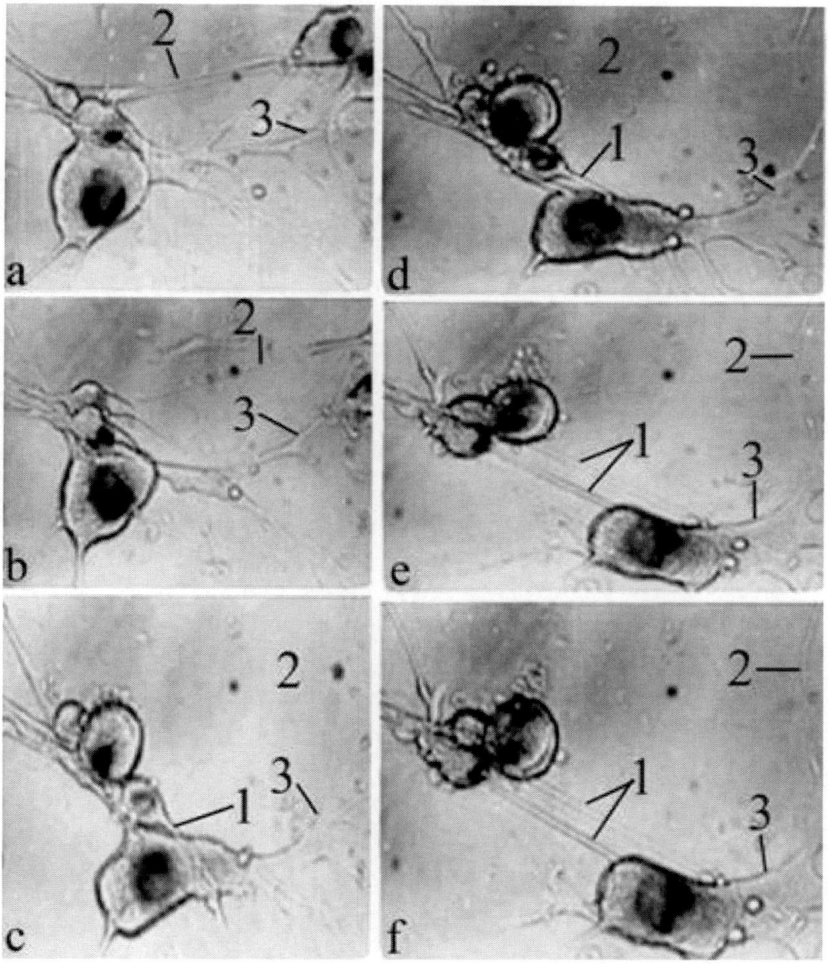

Figure 21. Extension and thinning processes (1) in tissue culture at two support points. The thinning of the nerve fiber down to submicroscopic visibility (2). Expansion of the dendrite (3). a-d - the dynamics of the process. Tissue culture. Intravital video microscopy. Phase contrast. Vol. 40 Ph, approx. 10.

Chapter 2

THE INVISIBLE FIBER

Anyone who has watched the amazing, almost transparent, "invisible" marine pelagic crustaceans at the biological station of the Academy of Sciences of the White Sea (Cape Kartezh), could not help but wonder at the beauty and mystery of these transparent animals, barely distinguishable in the surface layers of water. They live and camouflage themselves in the color of the water and the light. Also in other seas (Venus' girdle, the Yamato shrimp and others) (De Regge, Nauwynck, Geenen et al. 2006; Hanlon, 2007; Zylinski, How, Osorio et al. 2011; Feller, Cronin, 2014) (Figure 22). According to Kirchhoff's law (1861) every substance absorbs only those rays that it is able to emit under certain conditions (pelagic animals).

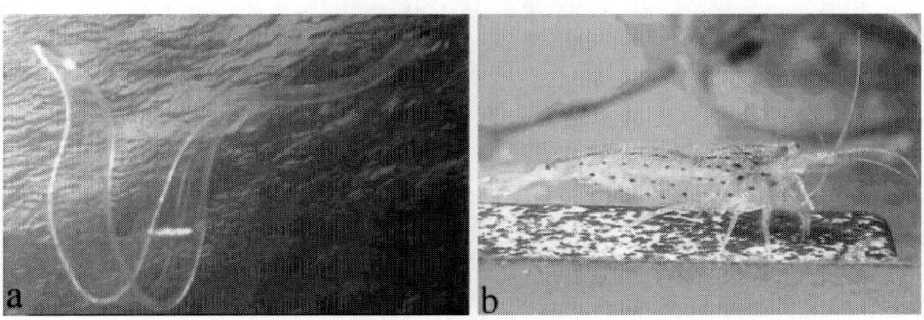

Figure 22. The varieties of camouflage of marine invertebrates. a – "Venus' girdle"; b – the translucent Yamato shrimp.

Perhaps they are visible only in the case of rapid short movements (Feller, Cronin, 2014). An observer might have called them "invisible animals," a phenomenon similar to that described in the book of Herbert Wells (1898),

"The Invisible Man." Something similar occurs in neuromorphology. Recently, in neurobiological literature the description of a surprising phenomenon was widely published. We are talking about the disappearance of the apical neural dendritic appendages under stress. Pyramidal neurons in the hippocampus, the prefrontal cortex and other brain regions have the strange property in certain conditions of rapidly "losing" their dendritic appendages, which subsequently reappear within two hours when former conditions are restored.

Authors usually note the rapid temporary loss and speedy recovery of the apical dendrite branches in many other emergencies (Popov, Bocharova, 1992; Popov, Bocharova, Bragin, 1992; Gulyaeva, 2013), such as during hibernation and re-warming in a number of animals. This surprising phenomenon has been described in vertebrates under repeated stress (Eiland, Ramroop, Hill et al. 2012; Kilinc, Gallo, Barbee, 2009; Malkinson, Spira, 2010; Miller, Morrison, McEwen, 2012) and in tissue culture (Fenoglio, Brunson, Baram, 2006). When creating "soft stress" through the introduction of corticosterone, a decrease in the length of the apical dendrites of the prefrontal cortex of 22-35% was observed. This phenomenon has been described by many authors in groups of rats under stress (Brown, Henning, Wellman, 2005; Grutzendler, Kasthuri, Gan, 2002). A rapid reversal of the phenomenon was shown under the effects of the anesthetic propofol (Martínez-Tellez, Gómez-Villalobos, Flores, 2005; Al-Jahdari, Saito, Nakano et al. 2006).

The unusual, significant, reversible shortening of dendrites and dendritic spines during hibernation was detected in ground squirrels (Arendt, Stieler, Strykstra et al. 2003), the European hamster (Magarifios, Li, Gal Toth et al. 2011), Siberian ground squirrels and other animals (Popov, Bocharova, 1992; von der Ohe, Smith, Garner et al. 2006). According to these authors, the restoration of the apical dendrites while the animals warm up occurs within two hours, which seems incredible. Contrary to our theory, based on light microscopy (the Golgi method), not all apical dendrites "shorten," "atrophy" or "disappear."

The authors showed that the average number of initial contact with the moss-like fibers of hippocampal spinelets decreases by less than one-third. That is, more than two-thirds of the contacts are preserved. There is a 26% loss of the total surface of contacts in the sections. Consequently, 74% of the area of contact corresponds to the preserved dendrites, which means that most of the synaptic apical dendrites must be contiguous, although they cannot be detected under the light microscope by the Golgi method. This data allows us to look for an alternative explanation of the effect of the "disappearance" of the apical dendrites.

However, a description of such mechanisms, which would provide for the "regrowth" within two hours or elongation of dendrites that had previously disappeared, does not exist. The assumption of a partial degeneration of dendrites is not tenable, because it does not leave behind any traces in the form of varicose deformation or pearl condition. Fibers simply disappeared from view, recalling visual fusion (Sotnikov, Malashko, Rybakova, 2006). It was assumed that the apical appendages were amputated, reabsorbed or atrophied. The issue of the appearance of transparency properties of animals was even discussed. However, this proved impossible (Kirchhoff, 1861).

No growth cones or signs of recovery are observed. This phenomenon seems to be impossible, but as shown by experiments on live fibers, it really exists. It is an indisputable fact, morphological mystery, and it should be explained somehow. However, on the preparations studied by the authors no signs of amputation resorption or atrophy were detected. But we did note however, the phenomenon of the retractile activity of axons. Perhaps the previous experiments with living nerve appendages will provide a solution to the problem.

Our attention should be given to the new phenomenon described by us (isometric retraction). Perhaps this is the only phenomenon that explains the thinning of the diameter of the axon and its disappearance, while all other observations of the appendages are limited to the shortening or reduction of their length. Thinning axoplasm occurs only under the influence of the retraction that happens when there are two opposing points of support, and can greatly change the diameter of the axon.

It was discovered that not only is there a reduction in the length of the apical dendritic branches, but thinning also occurs, an overall decrease in dendritic material (the volume of cytoplasm of apical dendrites) (Cook, Wellman, 2004). It had previously been noted (Sotnikov, Vasyagina, Sergeeva. 2013) that the nerve appendages, as well as retrograde cytoplasmic flows prompting a shortening of fiber, had a lateral displacement of neuroplasm and an increase of transverse diameter and volume when the length underwent changes. Under the chronic administration of corticosterone, dendritic material located proximal to the soma increased by 21% (von der Ohe, Smith, Garner et al. 2006). Important experiments were conducted by CS Cook and C.L. Wellman (2004). They were able to show, in some cases, an increase in their proximal dendritic material along with the reduction of apical dendrite branches under stress. This indicates that in the dendrites, as in the experimental axons, there is a simultaneous retrograde neuroplasmic flow, which increases proximal dendritic material (in the authors experiments to a maximum of 58%), possibly with the simultaneous reduction of apical dendrites. This suggests the movement of neuroplasm from the periphery of the arborization in a proximal direction (towards the cell body). The retrograde flow leads to a thinning of the basal dendrite branches. Unfortunately due to technical difficulties it is very hard to determine how much the diameter of a single dendrite increases upon retraction, since the appendage is constantly moving. It is known only from the experiment that the diameter of soma, upon the retraction of the dendrite into it, can increase in diameter up to 30% during 45 minutes of retraction (Sotnikov, Vasyagina, Sergeeva. 2013). It is also important to note that despite the substantial optically visible reduction in the amount of dendrites the number of synaptic terminals of axonal mossy fibers does not differ amongst the experimental groups (Magarifios, Li, Gal Toth et al. 2011). The proposition arises that under the described changes many inexplicably 'disappearing' fibers do not break off and do not regrow synapses anew as many think (Magarifios, Li, Gal Toth et al. 2011). Rarely thinning, they lose their argentophile property and cease to be visible. That dendrites don't break off is evidenced by the main characteristic of the phenomenon: the disappearance and extremely rapid restoration of the fibers is not possible in natural regeneration. The preservation of the 'seemingly disappearing' fibers upon hibernation is attested to by electron-microscopic preparations in which a multitude of preserved synaptic connections of the dendritic apical endings of the hippocampus and axons of moss-like fibers are visible (Popov, Bocharova, 1992). It begs for questions about the seemingly impossible proposition of the natural stretching and thinning of the dendrites. The experiments of D. Bray (1979, 1984) with artificially stretched nerve fibers in culture through use of a special micro-electrode up to 960 mcm over 24 hours are well known. The stretching of fibers and their 'tone' has been demonstrated by other authors (Roth, Bisbal, Brocard et al. 2012). The dynamic expansion of the 'visco-elastic nature' of axoplasm has been studied (Brown, Henning, Wellman, 2005) and its connections with the firmness of neurotubules. However, it seems that the phenomenon of the natural significant elastic expansion and thinning of nerve fibers has not been proven; but such a phenomenon has been discovered by us under conditions of isometric shortening. Consequently in this paper, our theory is that it is this phenomenon that is responsible for the mystery of the disappearance and re-appearance of apical dendrites in the experiments described above.

Experiments show that in the nervous system not only do the thin pre-terminal branches shorten and lose their optical density in tissue culture, but so do relatively large nerve stems. It turns out that living neural appendages display not only bi-directional motion but also a high natural elasticity[1], expressed by expansion. This multi-directionality is shown in experiments with living neurons.

We used cultures of neurons of peripharyngeal ganglions of molluscs, modified with ferments of pronase (Sotnikov, 2012). The retraction of these neurons continued for 2-18 hours, showing the dynamic of the behavior of appendages and the body of the cell. Any section of the nerve appendage is always accompanied by its slow and complete retraction (Figure 23), similar to the contraction of a strip of rubber. At the end of the appendage a more slowly contracting thin appendage (axial strand) is usually formed.

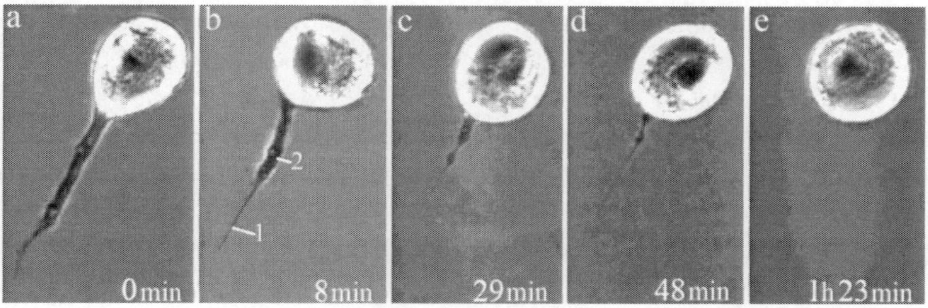

Figure 23. A neuron with a long non-shortening axial thin part of the fiber. a-d – stages of the process; 1 – the thin axial strand, coming out of the neurite; 2 – the retracting part of the axon. Time – from the start of observation. Intravital microscopy. Continuous video. Phase contrast. Vol. 40 Ph, approx. 10.

In other words, neuroplasm has the characteristic of elasticity. The speed of contraction of the preparation oscillates sharply (from 0.003 to14 mcm/min) amongst various fibers irrespective of their diameter. The direction of contraction of the fiber changes not just in regard to its place of attachment but also to the point of purchase (adhesion) along it. Naturally, contraction, thinning and lengthening of the appendage simultaneously means a concomitant movement of its neuroplasm mass. By changing the location of the point of attachment along the fiber, the direction of movement of the neuroplasm can also be changed (Sotnikov, Vasyagina, Sergeeva, 2014 a, b).

It is natural that the appendage itself will stretch and thin out more, the longer the process continues. Over various ten-minute intervals, the nerve appendage in the tissue culture may thin five- to ten-fold. An effect of this

[1] Elasticity presupposes light flexibility and expansion of material in a fully reversible process. Plasticity is the correct term for slight changes in the flexibility of material, which partially retains its new changed form (Kuznetsov, BTS 2002).

self-stretching is that the thinning requires there to be a general retraction of neuroplasm with the presence of two points of support. It is possible that this natural method of thinning normal fiber is a new way of changing its geometry and consequently its passive electrical properties. Since all the neurites in the tissue culture and the whole brain always have two points of support, isometric retraction may be considered to be unique and at the same time the most widely applied method of retracting appendages. We feel that it is possible to ascribe the quality of being the likely mechanism of the elastic thinning of nerve appendages to this phenomenon of isometric voluminal retraction, explaining the loss of their visibility (visual fusion). This proposition has been tested by us in experiments on tissue culture.

The loss of visibility of an object, as is well known, happens not only as a result of particular optical properties but also in connection with a reduction in its size and contrast. It is common knowledge that under a light microscope many morphological structures that are easily visualised under an electron microscope are not visible, even at the beginning of electron microscopic studies.

H. Fernändez-Moran (1952) described sub-microscopic nerve fibers invisible to a light microscope with a diameter of 0.1 to 0.06 mcm. This means that many morphological processes of living structures must also be invisible to light optical apparatus. Unfortunately such ultrastructural studies of disappearing apical dendrites are absent in the literature. But there is still one slim theoretical possibility which no one up to now has investigated. This is the unique rapid thinning of appendages to ultrastructural dimensions upon their natural elastic stretching. To what degree this is likely we will try to investigate in our present work. Nearly all the studies mentioned on the disappearance of apical dendrites are full of light optical methods. Consequently it's not surprising that the thinning of nerve fibers to ultrastructural dimensions is invisible. On several electron microscopic images of the axo-spinelets of synapses of the hippocampus upon hibernation (Popov, Bocharova, Bragin, 1992) there are groups of very thin (less than 0.1 mcm) fibers that are naturally invisible to a light microscope. One can conjecture that apical dendrites are capable of reaching similar dimensions upon retraction.

In order to confirm the experiments carried out on our isolated preparations, cultures of neurons of molluscs were studied (Figure 24).

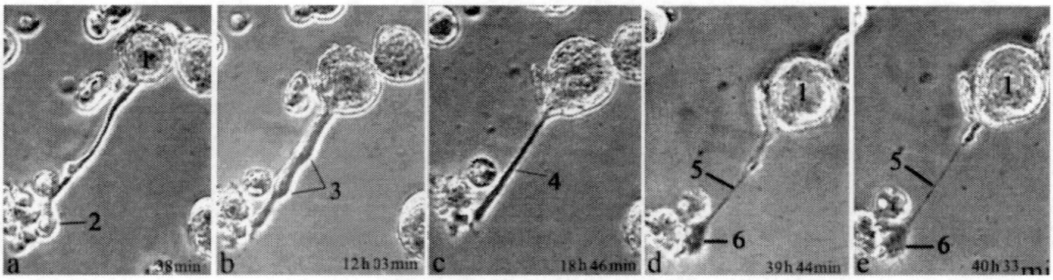

Figure 24. Changes in the diameter of fiber under retraction. The forming of isometric fiber. a – the original diameter of the fiber; б – thinning of the diameter of the fiber; в – thinning of the fiber as a result of the movement of the fiber into the body of the neuron; г – a sharp thinning of the fiber as a result of an increase in the volume of both ends of the preparation; д – the isometric reaction as a result of the sharp thinning of the axon between two points of support and its transformation into 'invisible fiber.' 1 – soma of the neuron; 2 – retraction bulbs at the start of the process; 3 – widening of the nerve fiber as a result of the reduction in retraction bulbs; 4 – sharp thinning of the fiber as a result of its retraction; 5 – isometric contraction. The forming of sub-microscopic fibers (invisible fibers); 6 – the sharp increase in the mass of the retraction bulbs as a result of the thinning of the material of the central section of the axoplasm. Intravital videomicroscopy. Phase contrast. Vol. 40 Ph, approx. 10.

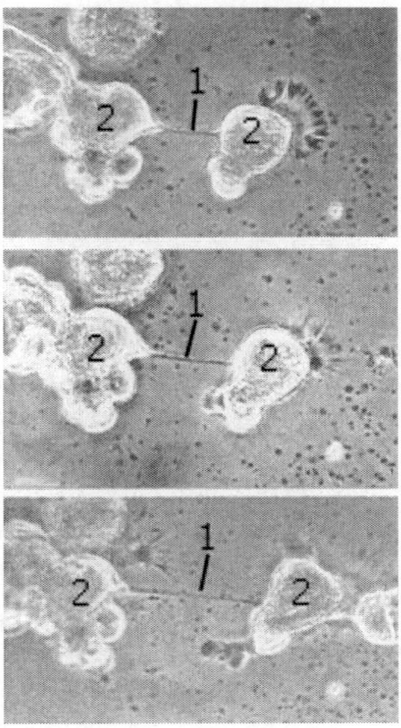

Figure 25. Isometric thinning and lengthening of the axon in opposite directions in tissue culture. 1 – the thinning and lengthening axon; 2 – groups of neurons moving in opposite directions. Intravital micro-videoscopy. Phase contrast. Vol. Ph 40, approx. 10.

On rare occasions one encounters another form of the thinning appendage. In these cases there are still two points of support. Thinning and lengthening of the appendage happens as a result of the dispersal of aggregate of the neighboring cells in opposite directions (Figure 25).

Maximum thinning occurs at the lamella of the regenerating neurons. The axoplasm spreads out around the body and the appendages. Before creating appendages it forms into thin flat lamellae (Figure 26).

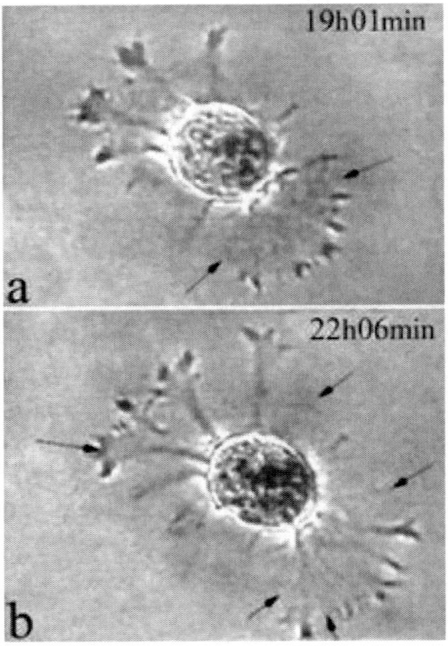

Figure 26. Lamellas – invisible in tissue culture. a,b – optically invisible lamellas, connected to cones of growth (arrows). Time – dynamic of regeneration. Intravital videomicroscopy. Phase contrast. Vol. Ph 40, approx. 10.

Invisible fibers are able to create before-light-optical plexuses, which appear unexpectedly in isolated areas of the regenerating neuron (Figure 27).

The diameters of the plexi are determined automatically in pixels. The mass of axoplasm moves along the nerve branches upon contraction. Consequently the movement of the object and the translocation was cultivated within five days. In the nerve plexi, it is convenient to analyze the dynamics of single fibers. The displacement of the mass of the neuroplasm was determined by the change in the thickness of the appendage. The disappearance of the appendage was judged to be at the point where the optical density of the preparation was equal to the density of the background (Figure 28). The numbers on the video frames were viewed by us as different symbols representing one and the same process.

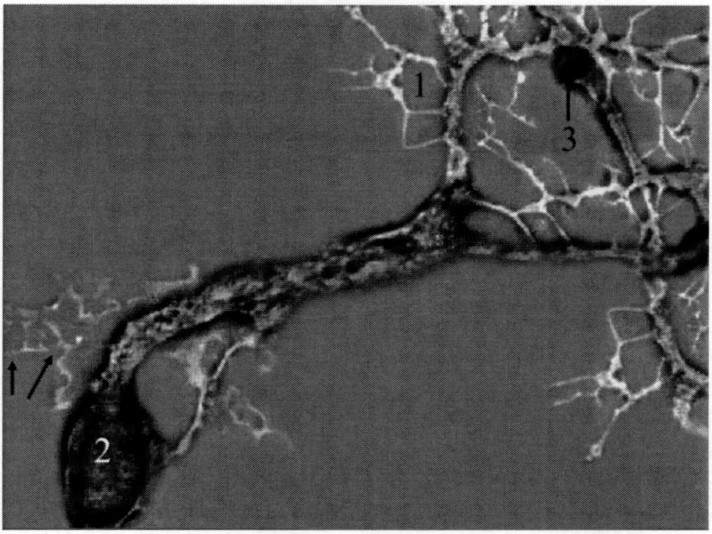

Figure 27. Before-light-optical plexuses, forming in the tissue culture (arrows). 1 – optically visible plexus in the tissue culture, appearing after 30 minutes; 2 – the body of the neuron; 3 – a synaptic preterminal in the tissue culture, reminiscent of a retraction bulb.

The fiber of the forming cultural plexi is fully comparable with the isolated fibers of molluscs. They also have a minimum of two points of support (the body of the cell and the synapse) and, interacting with each other, they form many such points. However, the retraction of these fibers in cultural plexi substantially differs from the contraction of single conductors. They demonstrate isometric contraction, changing the thickness of their diameter. The movement of their neuroplasm constantly changes in the opposite direction. In fact the retraction of fibers of the plexi in culture represents a special form of retraction. Some of the connected fibers thin out to such a degree that they cease to be visible through a light microscope (as though disappearing altogether) but then quickly restore their diameter and visible state.

So in Figure 28 a, the upper fiber (a_1) is widened with a significant increase in diameter. Fiber ($a_{3,\ 4}$) has almost completely disappeared from view[2]. As a result of the translocation of neuroplasm, its optical density is equal to the optical density of the background.

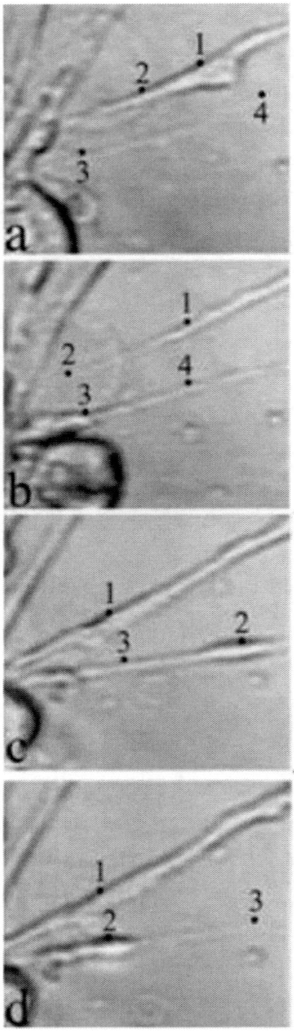

a

profile	fiber width	the density of the liber loop
1	21,00	133,00
2	10,00	114,00
3	9,00	99,00
4	0,00	0,00

b

profile	fiber width	the density of the liber loop
1	7,00	99,00
2	0,00	0,00
3	19,00	140,00
4	8,00	94,00

c

profile	fiber width	the density of the liber loop
1	15,00	155,00
2	18,00	177,50
3	12,00	107,00

d

profile	fiber width	the density of the liber loop
1	16,00	146,00
2	18,00	219,00
3	0,00	0,00

Figure 28. Dynamic of the change in thickness and optical density of nerve fiber plexi in a culture of dissociated neurons. a-d – dynamic of the process; 1-4 – Defining points of the diameter and maximum optical density of the circuits of the fiber (details in the text). Intravital video microscopy. Phase contrast. Vol. 40 Ph, approx. 10.

[2] It is important to iterate that the fiber has been placed on level glass with a stationary focus on the preparation.

To the contrary, on Figure 28 $b_{3,4}$, the fiber becomes clearly visible. The upper branch temporarily loses optical density, the fiber thins and disappears. The diameters of both fibers are restored after a few ten minute intervals (c_{1-3}). That is to say the process is readily reversible. On Figure d_1 the upper fiber has a high optical density whereas the lower fiber (d_2) again thins to such a degree that it loses optical visibility and becomes an 'invisible fiber'(d_3). In the central sections of the plexi, the maximal loss of optical density, the thinning of the fibers and their regeneration are all rule-governed processes.

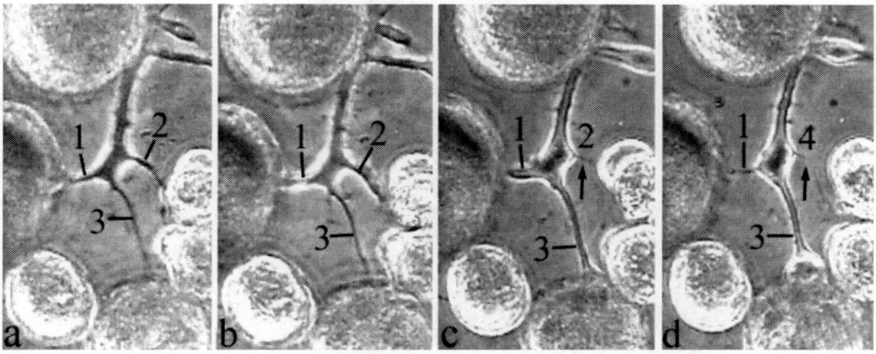

Figure 29. Dynamic of the volume of isometric retraction of inter-neuronal appendages in tissue culture. a-d – dynamic of the processes; 1-3 – preterminal nerve appendages constantly changing their thickness without a change in length; 4 – possibly, invisible fiber preserving the angle at its base (arrow). See text for a detailed description. Intravital serial video microscopy. Phase contrast. Vol. 40 Ph, approx. 10.

In Figure 29 it can be seen that one and the same pre-terminal fibers are constantly changing their thickness. Their diameter either expands (a_1-c_1), or thins to the limits of visibility (c_1-d_1). The thinned appendage (a_2) also either expands ($б_2$), or on the contrary becomes almost invisible (c_2). The same process of changing diameter is also observed on the fiber a_3-d_3. It is highly likely that the successive loss of visibility depends on a threshold level of thinning and loss of the optical density of the nerve appendages. Consequently, in studies of the dynamics of living neurons, as with other morphological objects, there is a significant optical interval that is rarely given attention, but perhaps studies would encounter the disappearing apical dendrites of the hippocampus. In other words, there is an objective obstacle, which prevents the full investigation of the mechanism of morphological kinetic states, in part, the motion of thinning dendrites upon their isometric contraction.

Local optical changes in the volume of living nerve appendages over short distances may transform them into 'invisible fibers,' and this process in the culture of neurons happens constantly. Appendages of the cells of neuroblastoma C-1300 Nirenberg (Figure 30) behave in exactly the same way. It is not just the fact of the multiple disappearance and the repeated appearance of the fiber but the incredible accuracy of its restoration that draws the attention.

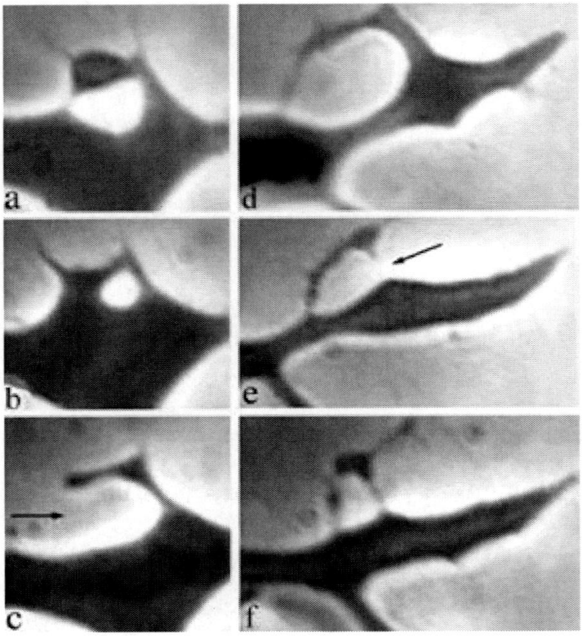

Figure 30. The repeated appearance of the invisible fiber in the area of the protoplasmatic ring of the appendage. a-f – stages of the process a, d, f – disappearing optical images of the fragments of the appendage; b – a full neuroplasmatic ring, c, e – invisible fibers (arrows). Intravital video microscopy (a culture of the cells of the neuroblastoma). Phase contrast. Vol. 40 Ph, approx. 10.

Often a picture is formed where seemingly completely detached appendages quickly re-establish themselves in the same size and exactly at the same place from where they disappeared. In other words, one can in complete certainty state that the fiber did not break off but temporarily disappeared from the optical field and reappeared again. This is also connected with the bi-directional movement of neuroplasm, which regulates the size of appendages (Sotnikov, Vasyagina, Sergeeva, 2013).

It seems that the effect of the thinning of nerve fibers is a widespread function, occurring in both intact appendages and under reversible reactive processes (Figure 31), connected with external action (the pearl condition).

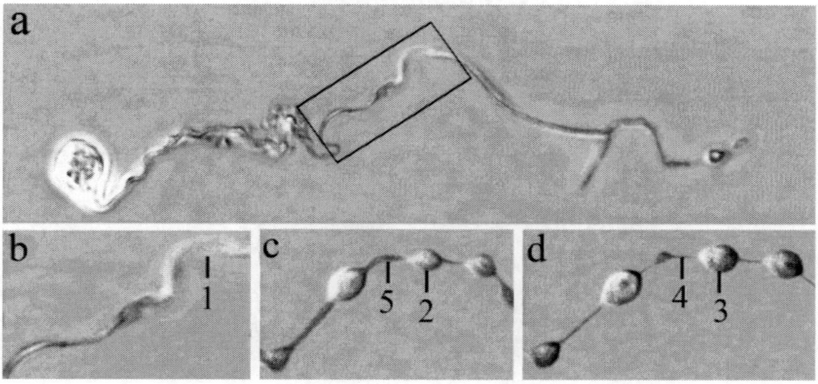

Figure 31. Sharp thinning of the nerve appendage in intervals between varicose areas upon phenomenon arising from the pearl state. a – preparation of isolated neurons with a preserved nerve appendage; b-d – dynamic of the process; 1 – original state of the fiber; 2 – appearance of varicosity as fiber undergoes stress; 3 – varicose rotundity; 4 – thinning of the fiber between varicose areas to the limits of visibility; 5 – local thinning of the appendage. The corresponding diagrams b–d. Intravital microscopy. Phase Contrast. Vol. 40 Ph, approx. 10.

The same phenomenon is observed in the formation of varicosity, beads of a moniliform state (Sotnikov, 1976; Malkinson, Spira, 2010). The passive electrical properties of the conductor (including the diameter) are essential for its electro-physiological properties. With the aid of time-lapse videography it has been shown that the varicosity of nerve appendages is not localized swelling, as many authors have suggested, but the dynamic redistribution of neuroplasm with sharp successive localized thinning and corresponding thickening of the neural conductors. Sharp thinning is also possible in the retracting fibers (Figure 32). Marked lopsided pearls are formed, seemingly according to the principle of Plato's unduloid (1876) with a constant increase in the curvature of the beads, their movement along the axon and the thinning of the intervals between them.

In tissue culture one often encounters the contraction of neurites, their convergence, the fusion of neurons and the formation of ganglions from single cells. However one also often encounters the spontaneous divergence of groups of cells. The appendages of the axoplasm display significant elastic properties however, and some appendages have thin to sub-microscopic dimensions.

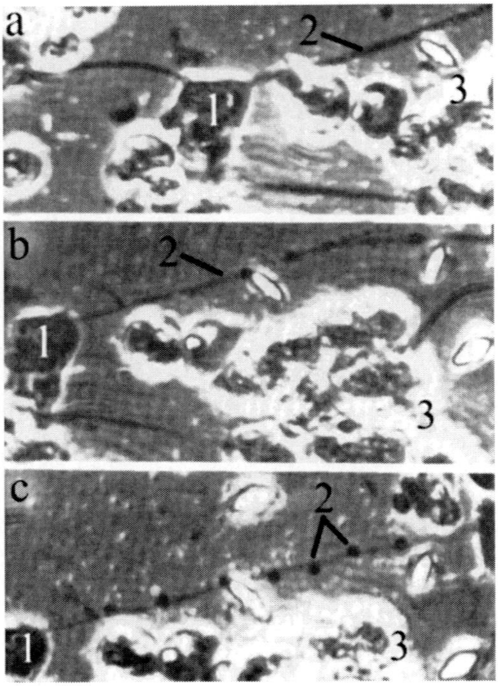

Figure 32. a-c – the body of a hydroid cell, modified with pronase; 1 – the moving hydroid nerve cell; 2 – the pearl state of the hydroid appendage by the Plato principle simultaneously with a contraction of the neuron; 3 – the remaining cells of the ganglion after the excision of the hydroid neuron. Intravital microscopy. Time-lapse videophotography. Phase contrast. Vol. 40 Ph, approx. 10.

The thinning and thickening may occur with simultaneous isometry (Figure 33). On image 34 we have depicted, with exceptional clarity, the size of the elastic thinning of cytoplasm of the nerve appendages. The diameters of the branches, in the given case, have thinned many dozen fold. It can't be ruled out that more pronounced thinning of the appendages has occurred with the loss of optical visibility. Whilst observing isometric contraction new conceptions of the mechanism of invisibility (transparency) of the thinning fiber and the kinetic state of the sub-microscopic process of the nerve fibers present themselves.

It is possible that the extremely extended 'invisible fibers' that have partially lost their affinity to the dye 'disappear' (become invisible). The fact that along the extremely emaciated fibers (as though broken off), are often located small particles arranged in rows, which provides evidence for the connections between these parts of the dendrites. They are not broken, just severely emaciated.

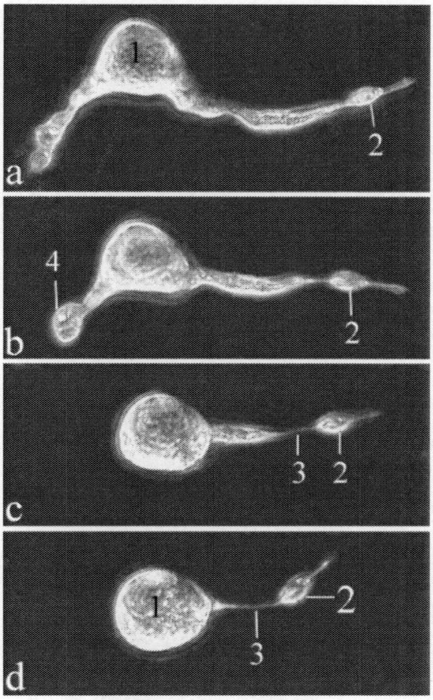

Figure 33. Sharp thinning of the appendage upon retraction of the axon. a-d dynamics of the process; 1 – original fiber; 2 – forming of varicosity; 3 – sharp thinning of the diaphragm between the varicose areas and the body of the neuron; 4 – retraction bulbs of the contracting appendage. Intravital microscopy. Serial videophotography. Phase contrast. Vol. 40 Ph, approx. 10.

We attempted to use data obtained from cultures of neurons and neuroblastoma to explain the 'reduction' of apical dendrites and the increase in volume of the proximal dendrites. The fully reversible fall in the optical density of the appearance of the nerve branches, losing their visibility, was viewed as the formation of mysterious 'invisible fibers.' We suggest that the incredible elasticity of neuroplasms and the special properties of isometric retraction are the true reasons behind the incredible phenomenon observed in the disappearance of the apical dendrites of the pyramid neurons. The reduction in the diameter of the fibers within limits of 0.1 mcm (Popov, Bocharova, 1992) may even change the chemical argentophilic properties of the fibrils involved in impregnation by the Golgi method, and contribute to the loss of visibility of the nerve appendages. Examples of the spontaneous fall of argentophile fibers are well known in neurohistology. They can also simulate a decrease in the length of the apical dendrites. Chemical and cytoskeletal changes of the neuroplasm during retraction of nerve appendages have been discussed a great deal in the literature (Baas, Ahmad, 2001; Myers, Tint, Nadar et al. 2006).

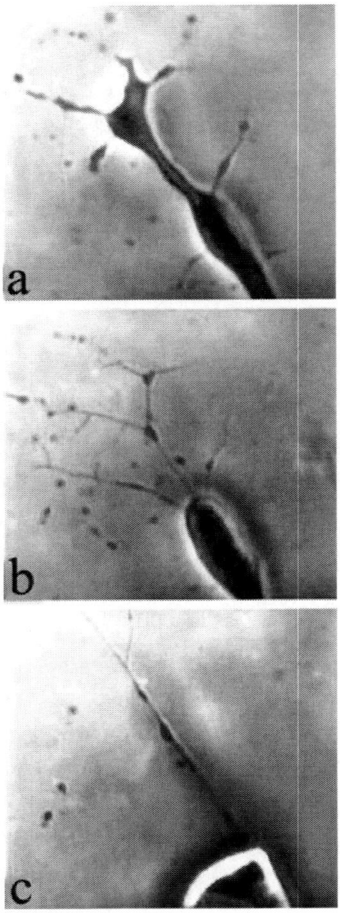

Figure 34. The sharp natural elastic stretching of neuroplasm upon isometric retraction. a-c – stages of the process; b-d – phases of the formation of invisible fibers. Their presence creates a greater quantity of rows of small dendrite varicosity along hypothetical straight lines; c – the optical disappearance of the thinning fibers. Intravital videomicroscopy (culture of cells of the neuroblastoma). Serial videophotography. Phase contrast. Vol. 40 Ph, approx. 10.

Dismantling of microtubules and their links with associated proteins is quite possible during cooling, as demonstrated by many authors (Arendt, Stieler, Strykstra et al. 2003). It was also assumed in this case that the ionic and osmotic homeostasis would be destroyed, an effect of the compound r190 (Billuart, Winter, Maresh et al. 2001). The disappearance of microtubules and neurofilaments were viewed as mechanisms of the rapid atrophy of dendrites under other experimental conditions. Recently, however, strong evidence has come to light (Baas, Ahmad, 2001) that the retraction is the result of acto-myosin contraction. In a hypothetical classification of morphological changes of nerve fibers, (Baas, Ahmad, 2001) suggested the possibility of thinning of the axons in the case of "thinning of the microtubule." It is more likely however that this is connected to the reduction in the intervals between the organelles and their clustering.

The above-described behavior of the nerve fibers in culture is consistent with isometric retraction. The first match is that the fibers must have two support points (both fixed ends). The second feature of this state is that fibers, contracting isometrically, possess surprising extensibility (several orders of magnitude). It is suggested that the phosphorylation of light myosin chains activate the acto-myosin processes in the axoplasm and can lead to "tremendous flexibility" of the regulating morphology of the axon (Baas, Ahmad, 2001).

Retracting neurons in tissue culture often leave behind a long, thin, trailing contractile residue. Apparently, appendages and some other appendage cells have the same property of abrupt elongation and thinning. For example, this is how appendages, ("dendrites") dendritic cells of lymphoid organs in tissue culture, behave based on the area of adhesion (Swetman, Leverrier, Garg et al. 2002). That is, our hypothesis of isometric retraction is based on the special double-fixing of fiber ends in the bi-directional flow of neuroplasms and their extreme flexibility and extensibility.

It is important to take into account the basic characteristics of the different types of microscopy. There are a lot of new options and tweaks, striving to overcome the difficulties in intravital microscope imaging ultramicroscopic processes. It is assumed that by the separation of molecules, both in time and in space (Mudrakola, Zhang, 2009), it is possible to achieve even "super resolution" of individual microtubules in living axons. But as the literature shows, there are still plenty of unexplored light optical macro-processes, such as the disappearance of the apical dendrites of pyramidal cells that are waiting to be solved. There is a specific unexplored space in intravital microscopic processes. It is this gap, in our opinion, which explains those remaining "secret" morphological phenomena such as "invisible fibers." When the invisible structural detail of "That-Which-Can-Not-Be" seems to be absent, or "atrophied," it may actually be alive and functioning, and its anatomic continuity maintained.

Apparently, neuroplasmic flows of voluminal isometric retraction occur in dendrites during hibernation or other changes of state. The direction of neuronal flow when cooled becomes mainly cellulopetal. It increases the mass of the basal dendroplasm (Figure 35). In this case there is a possible increase in the volume of the soma of neurons. During the warming up process of hibernating animals (Popov, Bocharova, 1992; Popov, Bocharova, Bragin, 1992), apparently neuroplasm flow direction is reversed. The fact that the restoration of the structure of dendrites occurs very quickly, possibly indicates that their anatomical rupture was absent during the changes initially, but the preservation of the old invisible paths or tracks contributed to the rapid recovery of anterograde neuroplasm flow.

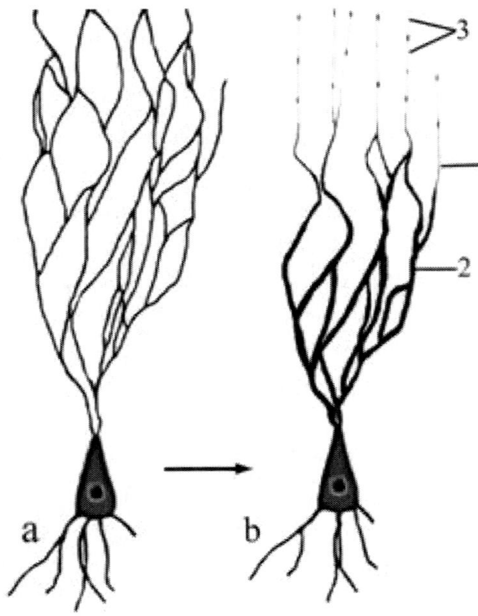

Figure 35. Hypothetical diagram, explaining the reversible "disappearance" of the apical dendrites of pyramidal neurons in their altered state. a - the initial state of the pyramidal neuron; b - morphological changes of the neuron; arrow - direction of the process; 1 - thinning of stretched apical dendrites; 2 - thickening of the basal dendrites; 3 – transformation of thinning dendrites into their invisible "croques" upon the alteration of neurons.

Emphasis is placed on the data demonstrating that under hibernation there is not simply a reduction in the length of the apical dendrites, but also retrograde displacement of the entire mass of the neuroplasm in the distal sections of the dendrites and their proximal cellulopetal translocation. In other words, one witnesses the same process of the repeated isometric retraction of the neuroplasm mass in whole neurons in vivo, similar to that which occurred in the experiments on the preparation of "nerve appendages of soma." The bi-directional voluminal displacement of neuroplasm in single fibers was detected (Sotnikov, Vasyagina, Sergeeva, 2013). We believe that the reduction in apical dendrites of hippocampal pyramids, the prefrontal cortex and the rapid restoration of the integrity of their structure after stress, hibernation, and other kinds of alteration is associated with the emergence and persistence of dramatically thinning "invisible fibers."

Thus, thanks to the study of living axoplasm we have been able to determine a new form of mobility that has allowed us to solve another mysterious morpho-physiological puzzle of invisible fibers.

Chapter 3

WHY ALL NERVE TERMINALS HAVE A SIMILAR STRUCTURE

As has long been noted by morphologists, bulb-shaped (sometimes large) protrusions always form in the area of injury of any nerve fibers,. This has been observed repeatedly Kamilo Golgi (1874), Ramon J Cajal (1959; 1995) and by many other classic neurohistologers both for diseases and injuries in the axons (Figure 36 a). However, it was also noted that such structures, which acquired the name "retraction bulbs," occur also in normal, and in latent pathology (Figure 36 b).

In numerous electron microscopic preparations, and in all schemas, the preterminals are usually depicted as retraction bulbs. Sometimes they reach a significant size, turning into a spherical formation ("Kugel-fenomenon" or the phenomenon of bullets), "incrustations" of axoplasm.

This form usually resembles a retraction bulb appearing after transection of the nerve fiber. In the past, many morphologists believed that such bulbs are somehow connected with the growth of axons, because they are often found in the nerve structure, where they are widespread, although they have different directions of growth.

It should be noted that retraction bulbs appear on traumas of the neurite at any distance from the cell body, or its terminals. From this it can be concluded that the phenomenon of retraction bulbs may arise at any level of the fiber - in both the synapses, and sensory terminals. Retraction bulbs occur at the extremities of normal single-terminal neurons with circular multilayer terminal coils in frogs (Figure 37).

In bipolar sensory neurons retraction bulbs repeatedly occur in normal sensory terminals. They are extremely reminiscent of the bulbs of efferent terminals (Figure 38).

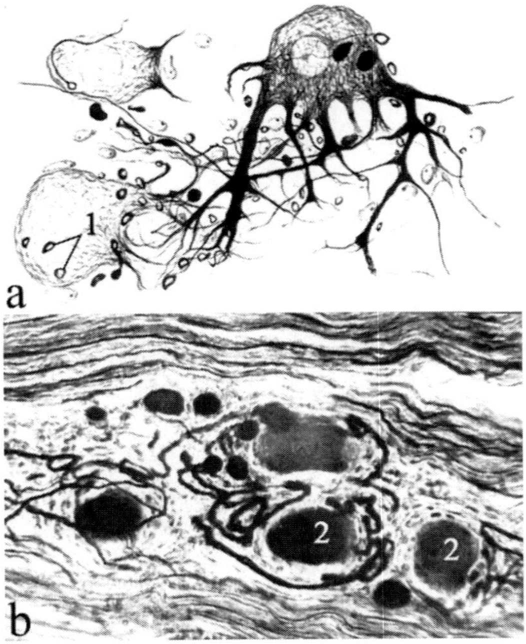

Figure 36. The bulb-shaped morphological changes of neuron terminals in latent and clear pathology. a - oval and bulb-shaped presynaptic terminals in the small intestine of a normal cat; b - retraction bulb of esophageal tuberculosis. 1 - presynaptic terminals; 2 - giant retraction bulbs of affected synaptic pre-terminals. Bielschowsky-Gros technique. Vol. 40 approx. 10.

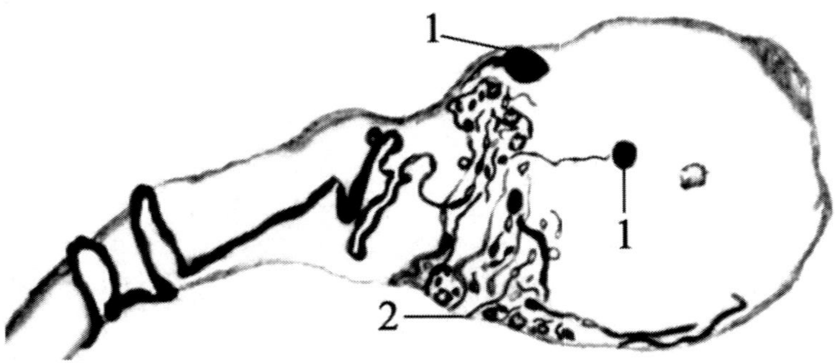

Figure 37. Axon and retraction bulbs in the area of synaptic terminal coils of the axo-somatic synapse of the normal unipolar neuron of the vagus nerve of a frog. 1 – bulbs of ultra traction in an intact unipolar neuron; 2 - normal axo-somatic synapses of synaptic coils of the neuron; 3 - preganglionic retraction bulb of the axon. Bielschowsky-Gros technique of silver impregnation. Vol. 40 Ph, approx. 10.

Such synapse-like formation can be seen also on free axonal neurites not closely connected with the pre-synapses. In fact, they can be regarded as signs of the dynamic processes of separation of the pre-synapse and post-synapse.

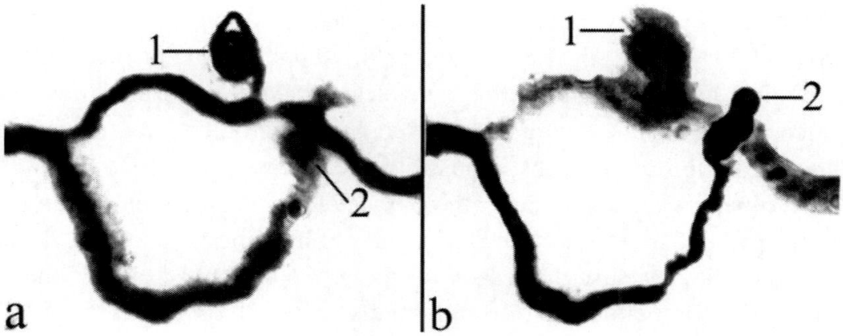

Figure 38. Single retraction bulbs of the endings of sensory neurons from the spinal ganglion of a cat.

Pronounced retraction of the sensory process in pathological states can dramatically change the shape of the cell body (Figure 39).

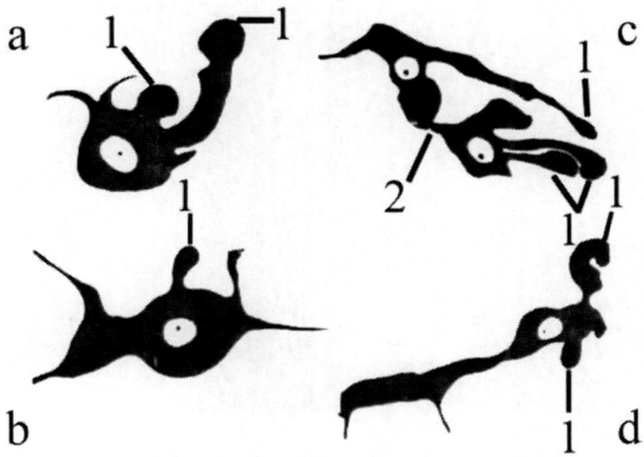

Figure 39. Retraction bulbs, deforming the body of neurons during intestinal obstruction. a-d - variants of deformation of neurons; 1 - retraction of the bulb; 2 - anastomosis between two modified neurons.

The history of the research of retraction bulbs is extremely interesting and long in duration. M.R. Duval (1895), and his team (Demoor, 1896; Stefanowska, 1897, 1898 and others), formulated the hypothesis that the synapses can delaminate. Pre-synapses by changing their shape, can be separated from post-synapses. The result is a physiological condition of sleep or anesthesia. The experiments were carried out on fixed preparations and therefore prompted a lot of controversy. In particular, the phenomenon was considered to be deeply pathological by our neurohistologer S.A. Soukhanoff (1898a; 1898b; 1899), but A.S. Dogel (1896) supported the original morpho-physiological hypothesis. Later this hypothesis was not confirmed, and it was forgotten. However, the emergence of free retraction bulbs in axons indicates that the separation of synapses is possible even when it comes to disease. At the same time, if we exclude the pathology, and only keep the initial retraction bulbs in mind, we cannot not go back to the old hypothesis of Duval, Demoor and Stefanowska, because the reason for the retraction bulbs, even at the very beginning, in our opinion, is the retractile tension of the synapse. We also know that rounding of the nerve ending reduces the area of its connection to the membrane of the post-synapse.

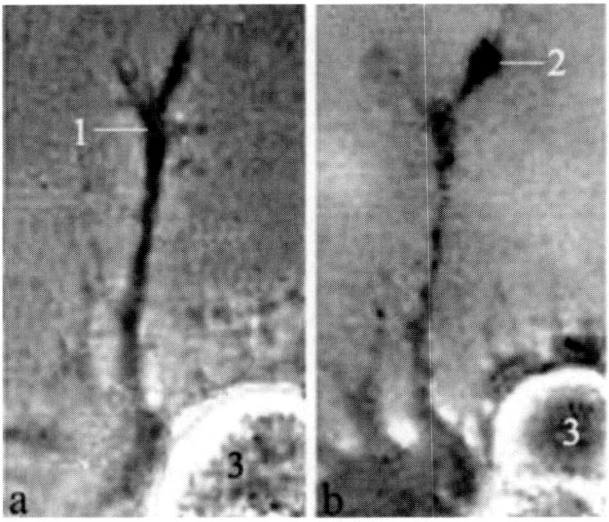

Figure 40. Dynamics of the transformation of normal growth cone into a retraction bulb in the development of nerve plexus in a culture of isolated neurons. 1 - the growth cone; 2 - retraction bulb of the same appendage; 3 - the body of the neuron. Intravital video microscopy. Phase contrast. Vol. 40 Ph, approx. 10.

The ability of retraction bulbs to form and disappear cannot be shown on fixed preparations. However, this phenomenon is easily and continuously demonstrated in tissue culture (Figure 40).

Retraction bulbs, forming from cones of growth in tissue culture, often form other cone-shaped structures (varicose), which, however, are not related to retraction bulbs and are not formed as a result of contraction. The growth cone can contain a large number of filopodia, which under retraction disappear first (Figure 41).

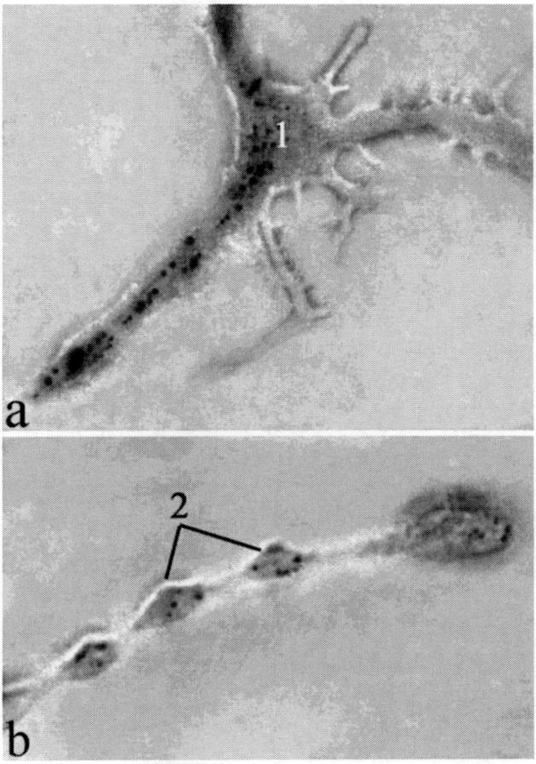

Figure 41. Growth cone (1) and retraction bulb of varicose fiber (2). Intravital investigation of light interference microscope. Vol. 100, approx. 15.

The terminal growth cone, which resembles maple leaves, losing filopodia, turns into a cone-shaped structure. The increasing growth cones, with little impediment, constantly and rapidly turn into retraction bulbs, demonstrating a growth dynamic. This process of changing the retraction bulbs and growth cones is repeated many times (Figure 42) and on different branches simultaneously.

This alternation of retraction bulbs and growth cones is a natural process of regeneration. Thus, the transformation of growth cones to retraction bulbs can be observed without any damage in natural conditions. Therefore it is always a question of the retractile capability of realizable retraction. Consequently, in the area of the ends of the cut fibers and in the area of the injured appendages and the natural growth cones a similar morphological pattern is observed, resembling a retraction bulb. Mechanical stress to the point of rupture of the axoplasm is easily traced in tissue culture.

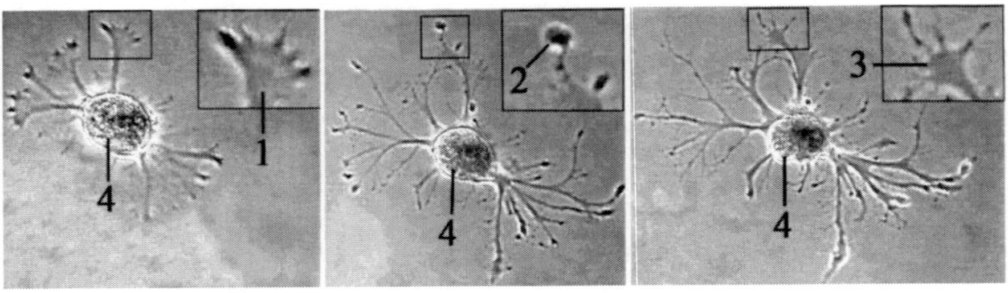

Figure 42. The dynamics of the transformation of normal growth cone into a retraction bulb in the development of nerve plexus in a culture of isolated neurons. 1 - growth cone; 2 - retraction bulb; 3 – repeated growth cone; 4 - the same neuron growing in tissue culture. Serial intravital video microscopy. Phase contrast. Vol. 40 Ph, approx. 10.

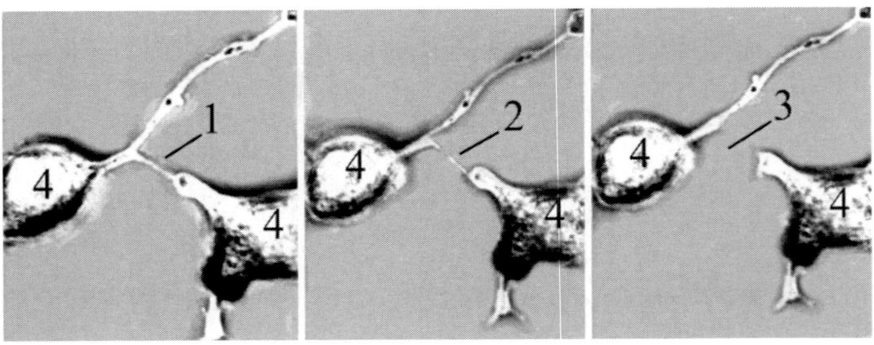

Figure 43. Natural retractile tension self-amputation of interneuronal synaptic contact and full retraction of the same preparation. 1 - the area of contact; 2 - tension of the preterminals before they break; 3 - rupture of the synaptic terminal; 4 – bodies of adjacent neurons. Serial Intravital video microscopy. Culture of isolated neurons. Phase contrast. Vol. 40 Ph, approx. 10.

Figure 41 captures the slow process of the separation of inter-neuronal contact. When watching the video film one can observe a slight strain, a thinning, and delayed retraction and contact between the two neurons. The pre-synapse then separates, and its terminals are reduced, breaking away from the adjacent fibers. In the study of living tissue of sensory terminals a similar phenomenon with the formation of retraction bulbs is frequently observed (Figure 43).

Retraction bulbs of sensory terminals in normal animals are frequently found not only in bush-like form, but also in encapsulated receptors with no apparent pathology. That is, under minimal external influence they may have multiple cone-shaped forms (Figure 44).

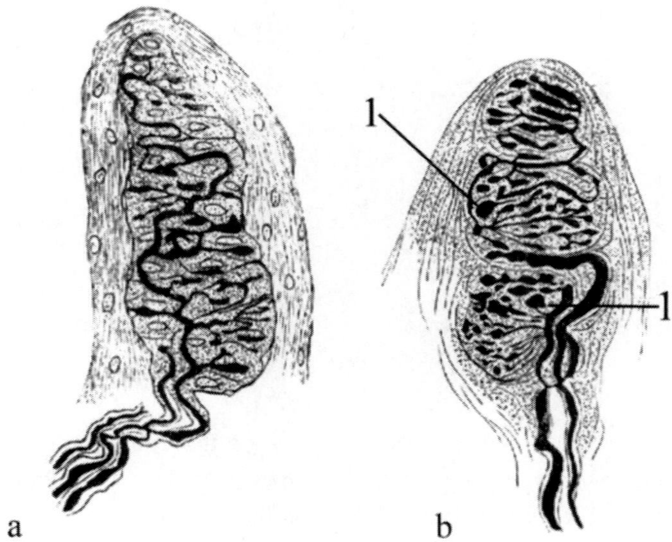

Figure 44. Variations of the structure of encapsulated sensory terminals. a - intact state of the receptor; b - multiple varicosity of the terminal. (A.S. Dogel, quot. At: Shtër, 1908).

Multiple vertical asynaptic dendrites (receptor analogues) of cells of the marginal layer of the cortex (Figure 45) have a constant multitude of bulb-shaped retractile structures on their terminals (Retzius, 1894).

Of great interest are the spinelet-like dendritic terminals with a bulb-shaped form (spinelet heads) and significant mobility. Their number and shape are volatile and are considered by many authors (Nimchinsky, Sabatini, Svoboda, 2002; Kasai, Matsuzaki, Noguchi et al. 2003; Knafo, Libersat, Barkai, 2005; Hao, Rapp, Leffler et al. 2006) to be related to the functional plasticity of the higher parts of the brain (Figure 47).

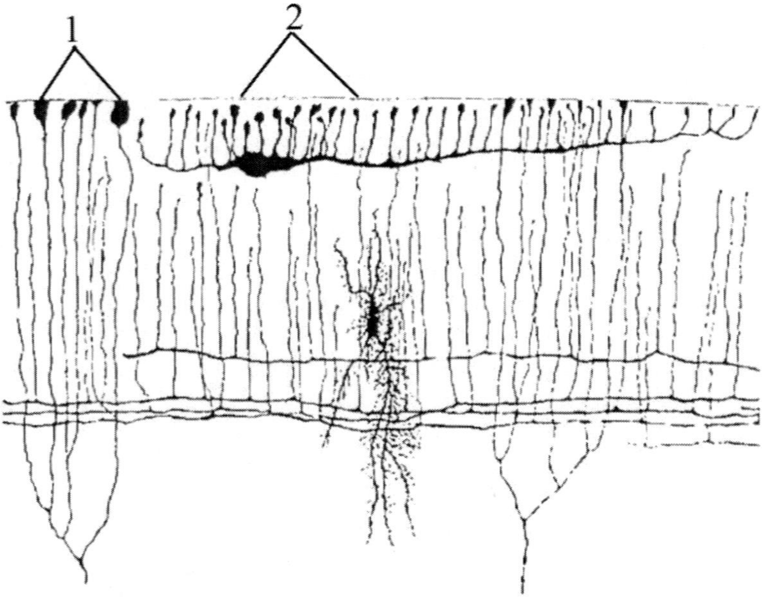

Figure 45. Retraction bulbs of asynaptic dendrites (resembling tissue receptors) of the cerebral cortex. 1 - retraction bulbs; 2 - retracted retraction bulbs (Ramon J Cajal, 1995).

Despite the physiological characteristics of the endings, their structural transformation is regularly repeated. The small leaf-like sensory patches, like growth cones, turn into spherical bulbs of the terminals of tissue receptors (Figure 46). Their endings suggest a similar retraction bulb as in synaptic nerve endings. Thus, contractile tone and contraction can be registered on fixed axonal and synaptic terminals, both intravitally in areas of interneuronal contact as well as in the axon injury near the soma of the neuron and the developing growth cones. The same signs of contractile tone are found in the study of sensory dendritic terminals. Similar bulb-shaped deformations in the spinelet-like dendritic terminals have been frequently described in literature of healthy and pathological states (Lukashin, Podolskaya, Soloviev, 1981).

We have repeatedly been able to observe the dynamics of the living tissue receptor endings in the transparent bladder wall of frogs. The picture was quite different than on fixed preparations. For clear identification of the terminals (of synaptic patches) a 0.001-0.02% solution of methylene blue is commonly used, cooked in Ringer's solution. At the beginning of the dye application and for a brief period, the receptor terminals look like a maple leaf with multiple pointed protrusive filopodia, quite unlike retraction bulbs, but exactly replicating small growth cones. Their movement is shown in Figure 46 b, d.

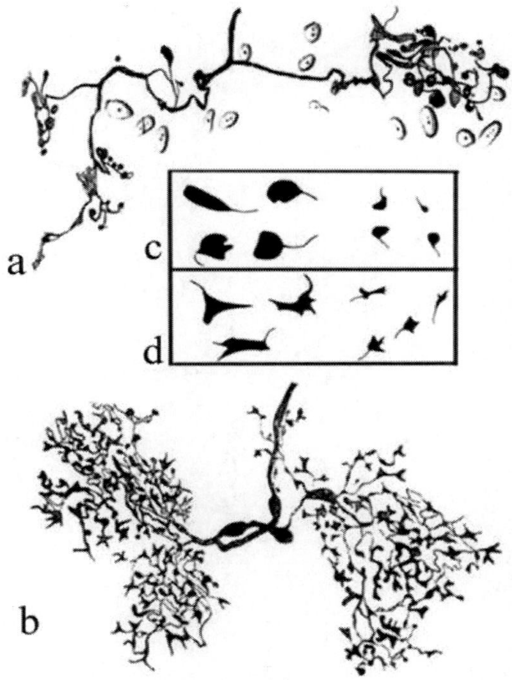

Figure 46. Retraction of sensory terminals of tissue receptors. a - retraction bulbs of the endings of intact tissue receptors, impregnated by the Bielschowsky-Gros technique; b - tissue receptors at the beginning of intravital staining with methylene blue; c - retraction bulbs of the terminals of asynaptic dendrites, fixed and impregnated by the Bielschowsky-Gros technique; d - sensory terminals of living asynaptic dendrites at the beginning of methylene blue staining. a,b - vol. 40 approx. 10; c, d - vol. 60 VI, approx. 10.

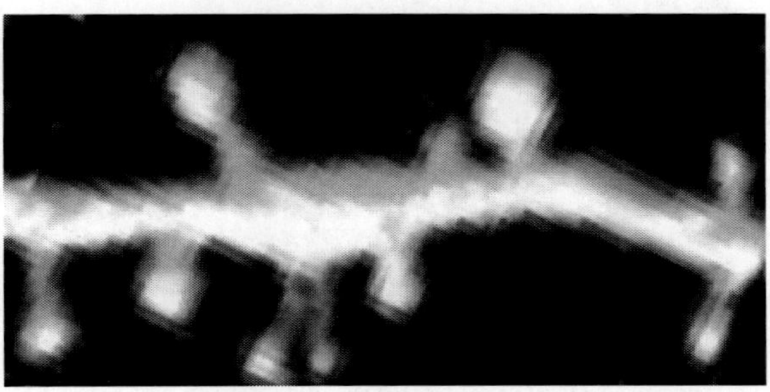

Figure 47. Spinelets of the hypothalamus with defined heads (negative).

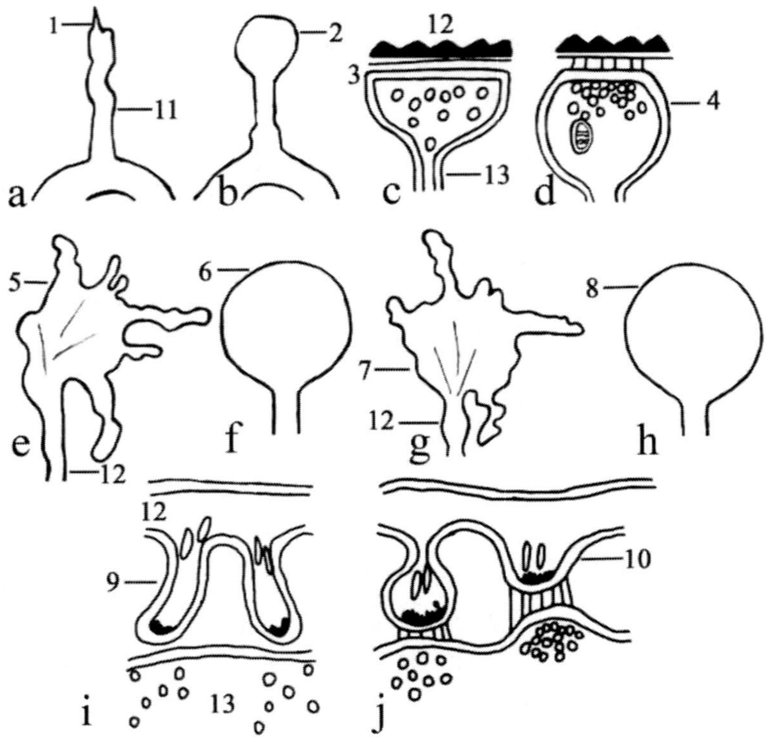

Figure 48. A schematic view showing the similarity of pictures of the retrograde tone of retraction of all types of terminals shoots. a-j – spontaneous retraction of various terminals of neurites; a, c, e, g, i – normal terminals in their original state; b, d, h, g, j – terminals after retraction (retraction bulbs); 1 –artificial ending of a fiber in the area of rupture; 3 – synaptic ending; 4 – bulb-like terminal of a retracting pre-synapse; 5 – a flattened growth cone; 6 – spontaneous retraction of the cone (retraction bulb);7 – flattened sensory terminal of tissue receptor in its original state; 8 – retracting terminal of sensory nerve ending; 9 – stretched form of spinelets in their original state; 10 – retraction of spinelets under changes in the homeostasis of their environment; 11 – neurites; 12 – dendrite or soma; 13 – axon.

A short time later, methylene blue, which many consider a satisfactory intravital dye, has an altering effect on the receptor tissue and the maple-like terminals. They round out and turn into structures no different from retraction bulbs. Thus, numerous varied patches of fixed preparations (tissue receptors) are converted into small retraction bulbs. If we assume that our observation is convincing and consider that it is formed as a result of retraction, it can be assumed that the process of modification and alteration of sensor terminals is the same as in other parts of the terminal sections of the neurite (Figure 48).

Thereby once again it is confirmed that the reason for the same type of structure of retraction bulbs is retractile tone. Since the growth cones and terminal sensory patches are identical to each other and are constantly in motion, it can be assumed that the end of the receptor tissue, not related to other parts of the neuron, can move independently at some distance and scan the innervated area.

Thus, a general idea of the form and reason for the similarity of nerve endings arises. They do not depend on the physiological functions performed by the neuron, either motor, sensory, or growth. In other words, there are certain physical laws, creating a single form, regardless of the functional state of the object and its properties (Figure 48). As is known, the highest surface tension (Gibbs' free energy) is possessed by bodies with a spherical surface. The end structures of nerve fibers strive towards the same form at any level of their organization. The experiments demonstrate the presence of retractile tension in all cases of alteration of the nerve process, including transection, pathology and natural physiological movement. All these neural structures end in similar bulbs due to the fact that they all possess the same retractile ability of axoplasm.

Therefore, any form of ending noted previously by neuro-morphologists, has similar causes. In fixed preparations, the end patches have bulb-shaped forms that develop in connection with the activities of stimuli, including methylene blue. This solves another secret of axoplasm.

This concludes the group of experiments on living axons, confirming that axoplasm has retractile properties and is able to perform unusual morphological and physiological tasks. Retractile propensity and retraction are perhaps the major non-electric features of axoplasm.

Chapter 4

Nerve Fiber Trauma and the Retraction of Axoplasm

As is known, the effect of the structural and physiological recovery of transected nerves depends on several fundamental processes. First of all, as the majority of authors note, this regenerative ability of the proximal segments of the fiber, which is largely dependent on the metabolic activity of the bodies of related neurons and of the state of anterograde axonal transport of nerve conductors. The most important prerequisite for successful regeneration of connective tissue is also the state of collagen in the area of the nerve rupture, which is heavily dependent on the magnitude of the gap between the proximal and distal segments of nerve (diastase). The rate of resorption of the products of the decrease in myelin of the destroyed membrane segments. All these components are taken into account in the treatment of damaged nerves[1].

Much attention is paid to the use of various neurotrophic factors (Kalyunov, 1986; Dolbeare, Houle, 2003; Klimaschewski, Hausott, Angelov, 2013). Other than nerve growth factor, neurite-stimulating brain proteins, soluble neurite-stimulating factor, ciliary neurotrophic factor, epithelial growth factors and hepatocytes were also examined. There was a fundamentally new method of activating regeneration by gene expression of proteins under electroporation of the neurons of the spinal ganglia (Lerch, Martínez-Ondaro, Bixby et al. 2014).

In addition to the regenerative ability of nerve transection in the treatment of this pathology an important place belongs to attempts to define the

---

[1] The large volume of work dedicated to methods of healing severed nerves is impossible to cite within the limits of one work, so we had to refer to a small number of articles published relatively recently.

differences between the stumps of the transected nerve and the forming connective scar. It was supposed that the appearance and size of the defect of the nerve was associated with the elasticity of the nerve membranes. A large number of original methods for the prevention and elimination of diastase were used. First of all, attempts were made to bring nerve stumps close together with a conventional seam of epineural membrane or microsurgical methods of stitching the perineurium of nerve bundles (Martins, Teodoro, Simplicio et al. 2011; Fraczek-Szczypta, 2014). For this purpose bio-resorbable suture material and biocompatible nanofibers are currently used (Fraczek-Szczypta, 2014). The alternative to nerve suture is biological glue, which was more effective than traditional stitches when connecting stumps (Gencer, Özkiriş, Saydam et al. 2014; Erfanian, Firouzi, Nabian et al. 2014). Efforts were also made to combat the growth of collagen in sutures and nerve neuromas (Turgut, Uysal, Pehlivan et al. 2005; Hellal, Hurtado, Ruschel et al. 2011; Shen, Wang, Gu, 2014). Attempts to prevent the emergence of diastase by stitching stump not "end-to-end" but "side-to-side" or "side-to-end" had no significant effect in increasing and accelerating recovery (Artiaco, Tos, Conforti et al. 2010; Radke, Allmeling, Waldmann et al. 2011).

From our review of modern literature it is clear that, despite the many already tested methods and techniques of treatment to prevent occurrence of post-operative diastase, and collagen scars and neuromas, it, is not yet completely possible. Is it possible to stop the discrepancy fibers forming diastasis or is it?

It would seem that the treatment of this disease requires new techniques based on other properties of nerve fibers. In particular, it has long been shown that severed axon cells in tissue culture, as well as any other appendages of living cells, are able to contract (Lowery, Vactor, 2009). The same is true of axons after injury. This means that upon nerve transection, the retraction of nerve fibers can participate in the process of the forming diastase. Thus, neither seams nor glues, which firmly join nerves, prevent the development of diastase, which almost inevitably occurs. It is possible that in addition to the degeneration of inter-nodal transected segments, the mechanism of formation of diastase includes the process of retraction of fibers. Perhaps diastasis occurs not as a result of passive mechanical, elastic shortening of axons, but as an effect of their contractile activity. Given this, the connective tissue of the nerve membranes should always be outside the distal and proximal axons, which determines the location of the fibrous scar, its own active and regenerating axons distant from each other.

Regeneration, as is known, does not begin immediately after the rupture of the nerve, but sometime later. This latency period is necessary for the deployment of a complex preparation of cellular and molecular mechanisms, preceding the beginning of the regeneration (Klimaschewski, Hausott, Angelov, 2013). It can last days, and it is sufficient for the growth of scar tissue. Unfortunately, the methods for combating the reduction of latency time of regeneration in the literature are clearly absent. Apparently, it requires special techniques to prevent retraction and divergence of the severed nerve stumps. We believe that, given the above, the possibility of inhibiting the retraction crossed nerve fibers in the experiment must be explored.

Experimental, theoretical and molecular biological studies of the retraction of axoplasm proteins are widely represented in the literature. It has been shown that the activity of the tubulin is blocked with colchicine, cytochalasin inhibits actin, myosin activity is blocked by blebbistatin, and the blocker of L-type calcium channels is nimodipine. Therefore, it was decided to use them to identify the molecular mechanism of reduction of axons and try to stop it, to prevent or reduce the expansion of diastase.

The material was obtained from the amputation of the lower leg in a specialized neurosurgeon Y.A. Shulev (2005), 3-4 days after mine blast injuries. Fragments of the n. tibialis (n = 9) were fixed in 12% neutral formalin for several months. Sections were impregnated by the Bielschowsky-Gros technique. Doing this, we had in mind previous experiments with translocation of axoplasm mass. First, we investigated signs of retraction of nerve fibers of the peripheral parts of the n. tibialis. As a control, we used fibers located 10-20 cm from the injury. The fibers, in sections of 30-40 microns thick, were impregnated with silver nitrate by the neurohistological procedure, the Bielschowsky-Gros technique. It was necessary to start the study on live myelinated fibers, as the possibility of their earlier retraction has not yet been studied, and the complex structure of their myelin membrane could provide an obstacle to regeneration.

In control preparations of amputated human nerves (Figure 49 a) intensively impregnated axons without any deformation, blistering, signs of damage, reduction and degeneration were found. On some of them weakly colored contours of myelin membranes with Golgi cones were detected in the area of the Schmidt-Lanterman section. Near the nerve section (Figure 49 b), in preparations of myelinated fibers, a large number of isolated ellipsoidal fragments of degenerating fibers and varicose deformation of axons, which are known as symptoms of non-specific reactive restructuring, were detected (Sotnikov, 1976). Between the pronounced varicose sections are areas of thin dense axial strands, which remain intact during almost complete demyelination and are destroyed last. Additionally, at the ends of the severed fibers were found characteristic swellings - retraction bulbs (Figure 50 a). They were previously known as "retraction bulbs," "kugel phenomena" such as are found in large quantities in the neuroma during nerve regeneration. In fixed preparations, fragments of damaged fibers are also found, which have such bulbs on both of the severed ends (Figure 50 b).

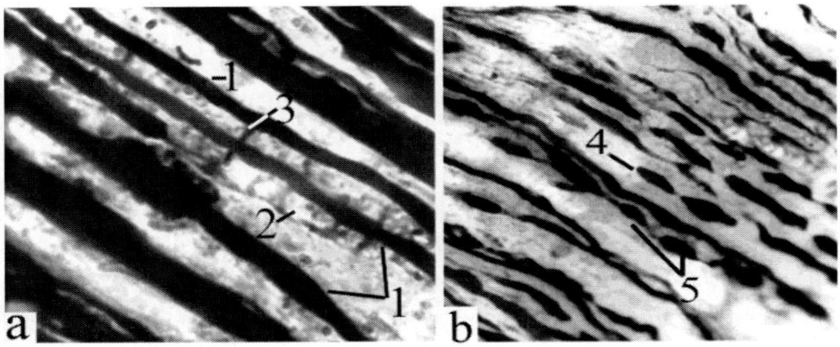

Figure 49. The difference between proximal and distal fibers of the amputated nerve stem injury. a - relatively intact axons; b - varicose-deformed myelinated fibers near the injury; 1 - axons; 2 - myelin membrane; 3 - Golgi cone in the severed area; 4 - degenerated nerve fibers 5 - varicose axons in the initial stage of reactive restructuring of fibers. The Bielschowsky-Gros technique. Vol. 40 approx. 10.

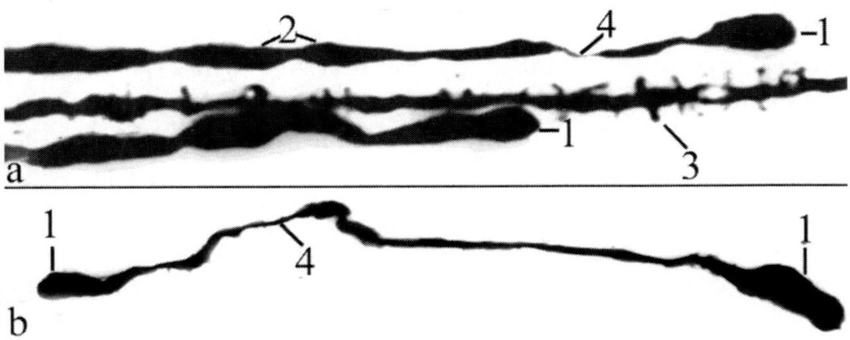

Figure 50. Indicators of the active contraction of severed nerve fibers, their early structural reaction and the complete degeneration in fixed preparations of human nerve after a mine blast injury. a – retraction bulbs of the severed fibers; б – retraction bulbs on the opposite ends of the severed fiber, corresponding with the dual-directionality of axonal retraction; 1 – retraction bulbs; 2 – varicose deformation; 3 – Golgi cone; 4 – axial strand. The Bielschowsky-Gros technique. Vol.40, approx. 10.

It became clear that the information about the damage of peripheral nerve fibers is passed through the body of the neuron by retrograde transport (Ben-Yakov, Dagan, Segal-Ruder et al. 2012). Such a rare illustration, nevertheless, confirms the bi-directionality of axoplasm "flow," as discussed above. The idea of end bulb-like growth cones at the two ends of the isolated fragment of the fiber is not possible. Based on previous experiments the phenomenon can be explained only by the appearance of retraction bulbs. The phenomenon described can only be explained by the retraction of fibers of both ends of the fragments of the severed nerve. The reality of the reverse flow of axoplasm of animals is already firmly established in science (Hasaka, Myers, Baas, 2004; Mann, Rougon, 2007).

However, the possibility of the contraction of the living myelin nerve fibers, which make up the majority of peripheral transverse nerves and pathways of the brain, has not been studied (Figure 51). The possibility of the retraction of conductors was investigated by us on single living myelin fibers of the sciatic nerve of the frog.

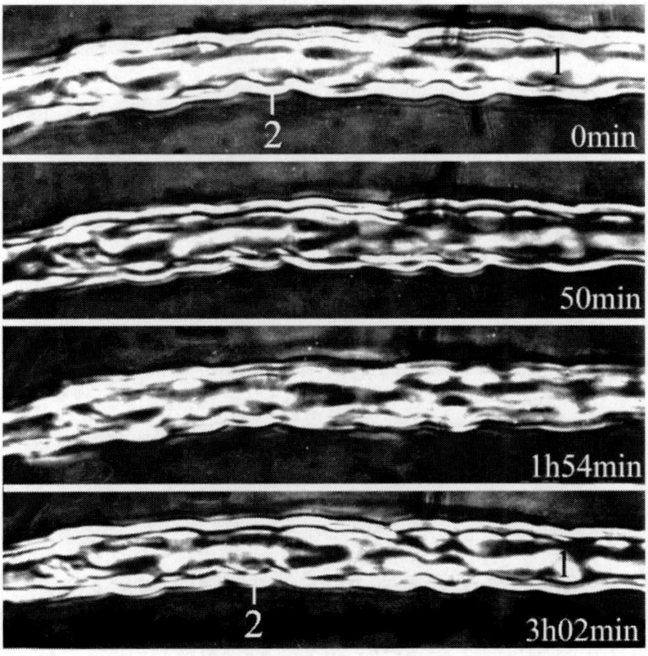

Figure 51. Contraction of the structure of living myelin nerve fiber under stress in Ringer's solution over a three-hour duration. a-d – time-lapse serial photography of the same fiber over various intervals of time from the start of the experiment; 1 – axon; 2 – myelinated membrane. Phase contrast. Vol. 40 Ph, approx. 10.

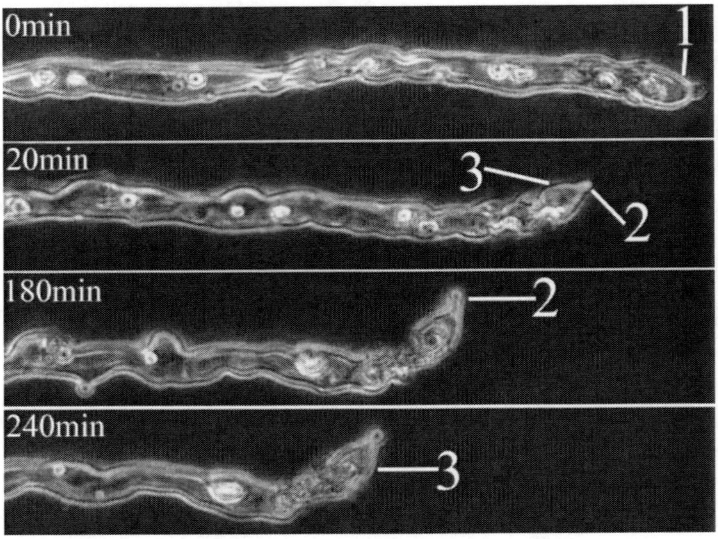

Figure 52. Retraction of a living myelinic nerve fiber after transection. 1 – sealing of the severed end of the fiber; 2 – formation of the axial strand; 3 – the forming retraction bulb; phase contrast. Intravital serial videophotography. Vol. 40 Ph, approx. 10.

To prove the contraction of living myelinic fibers and study this process, single fibers of a split sciatic nerve of the frog Rana temporaria was used. The fibers were studied in Ringer's solution using phase contrast video microscopy with an inverted microscope and special micro-camera. The contracting fibers were recorded every 10-15 minutes. Such isolated preparations can stand without undergoing changes in Ringer's solution for 2-4 hours and can be used in experiments. When the myelinic fibers were severed their contraction was indeed found (Figure 52).

The retraction was detected at both ends of the transected fragments. The rate of reduction varied greatly in different preparations from 0.007 to 2.7 m / min. Perhaps this is due to the different distances from the site of transection to the next intact node of Ranvier or the extent of damage of the test fiber. At the ends of the stumps a "sealing" of the regions of the rupture through the merger and the formation of myelin retraction bulbs occurs (Figure 52).

From the cut contracting end, a slower contracting tight axial cord appears, usually protruding from the retraction bulb (Figure 53). It's the same little-known central structure of the axial cylinder, which appears between varicose areas in fixed nerve preparations.

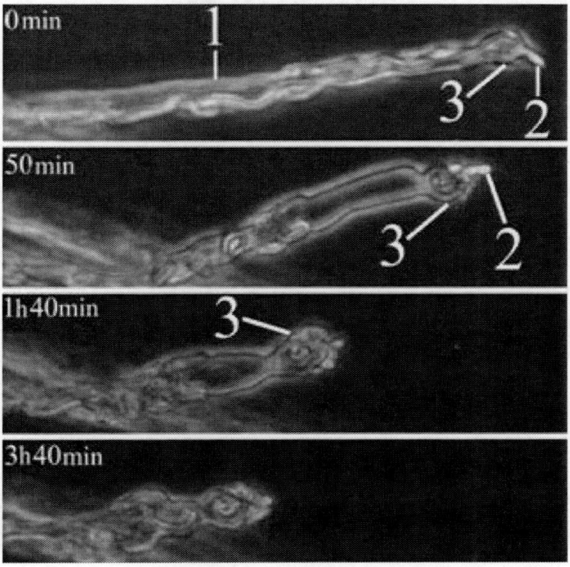

Figure 53. Remains of the axial strand in the area of the end of the severed inter-nodal segment. 1 – unchanged area of the fiber; 2 – axial strand; 3 – retraction bulb. Phase contrast. Intravital serial videophotography. Vol. 40 Ph, approx. 10.

Based on observations of the reduction of myelin fibers, it could be assumed that the retraction of transected axons is indeed involved in the formation of the interval between the nerve stumps. Moreover, this interval should be increased twice due to the bi-directional retraction of the segments. Apparently, because of this the nerve seam connecting the perineurium and epineurium stumps during the operation, in principle, cannot closely attach the nerves, because the contracting fibers diverge and not the nerve membranes, which are stitched together. It is to the area vacated by retracting the nerve fibers that fibrous tissue rushes during the post-operative period. It should also be noted that under transection of fiber the last to be damaged and degenerate is the Schwann cell. Demyelination of the membrane is limited to that nearest the Ranvier node, that is, the adjacent Schwann cell. The varicosity observed on the fibers of the damaged nerves of frogs and human, although associated with nerve injury, is not directly related to its contractile activity to our knowledge. Retracting fibers without varicosity apparently have the same speed of retraction. In the area of the cut end of the fiber the phenomenon of self-amputation is often observed (Figure 54).

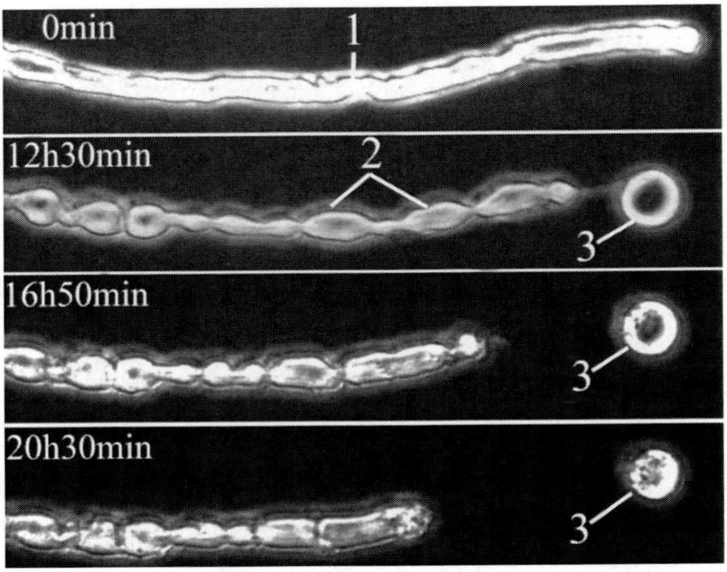

Figure 54. Self-amputation of the retraction bulbs of the transected myelinic nerve fiber upon contraction. 1 – unchanged area of the fiber; 2 – varicose deformation of the same fiber; 3 – amputated varicose fragment. Phase contrast. Intravital serial video photography. Vol. 40 Ph, approx. 10.

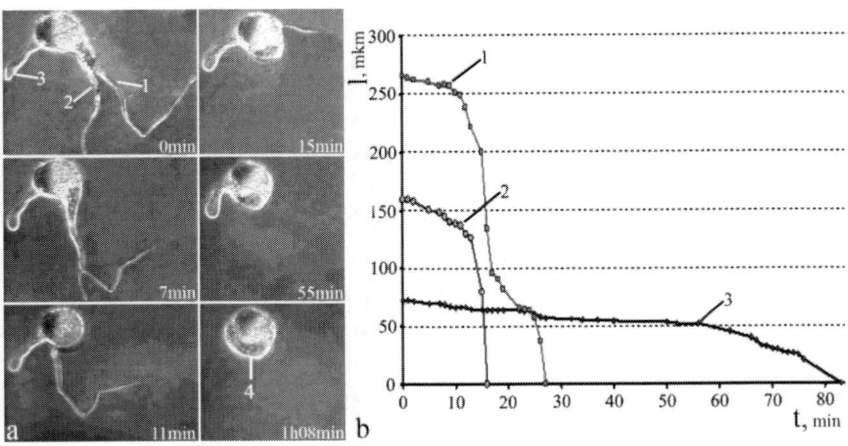

Figure 55. Different rate of retraction of various appendages of the same neuron. a – dynamic of the process; b – graph of the changes in the rate of contraction of various appendages of the same neuron; 1-3 – identification of various appendages and their rates of retraction; 4 – a neuron with a fully retracted appendage. Phase contrast. Serial videophotography (a), graph (b). Vol. 40 Ph, approx. 10.

We have tried to use data on the contractile properties of living axons, information from electron microscopy and from molecular biology on the effect of inhibitors to stop the contraction of injured axons, trying to reduce the amount of diastase. For this, we studied the therapeutic effect of colchicine (blocker of the activity of tubulin), cytochalasin (inhibitor of the activity of actin), blebbistatin (stabilizer of the movement of myosin) and the calcium channel blocker - nimodipine. We also proposed to reveal the toxicity of these drugs when used for medicinal purposes. For this electrophysiological microelectrode experiments were conducted on the single neurons of leeches.

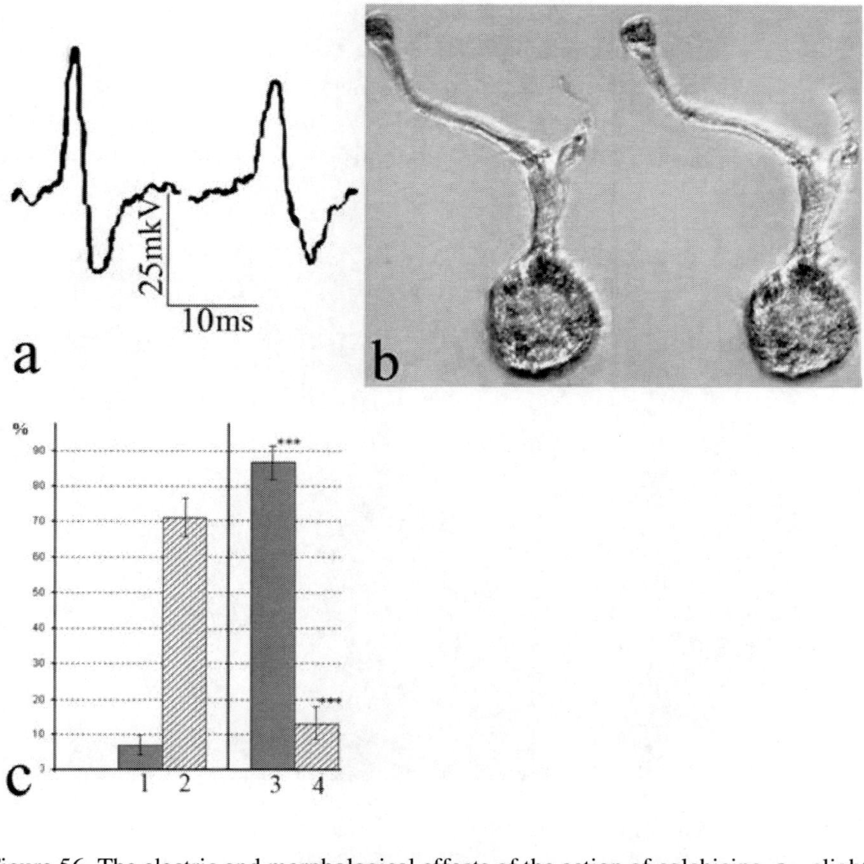

Figure 56. The electric and morphological effects of the action of colchicine. a – slight drop in the potential of the activity of the neurite; b – absence of the retraction of nerve fibers; c – percentage of content not contracted in the colchicine (1, 3) and contracted (2, 4) in the control group of nerve fibers. Note: *** - reliable significant variations from the control at a level of significance r ≤ 0,005.

Figure 55 shows the dynamic of the contraction of three appendages of a living neuron in control setting, to demonstrate the actual retraction upon the use of therapeutic drugs. It is shown that the rates of retraction of appendages of the same neuron are very different. The shortest axon can be invaginated into the body the longest, that is, to have the lowest rate of retraction (Figure 55 b). In the control a natural spontaneous contraction was observed in 82.4% of neurons.

The data gained on the participation of motor proteins of the cytoskeleton in the mechanism of traumatic retraction seems natural. However, the role of tubulin in the microtubules in the mechanism of the contraction of fibers during trauma requires additional studies. The hypothetical notion of the participation of changing microtubules in the contractile activity of fibers is known. Therefore, we considered it necessary to determine the possible involvement of tubulin microtubules also in traumatic retraction. To do this, we used a solution of colchicine, commonly used for the dissociation of microtubules.

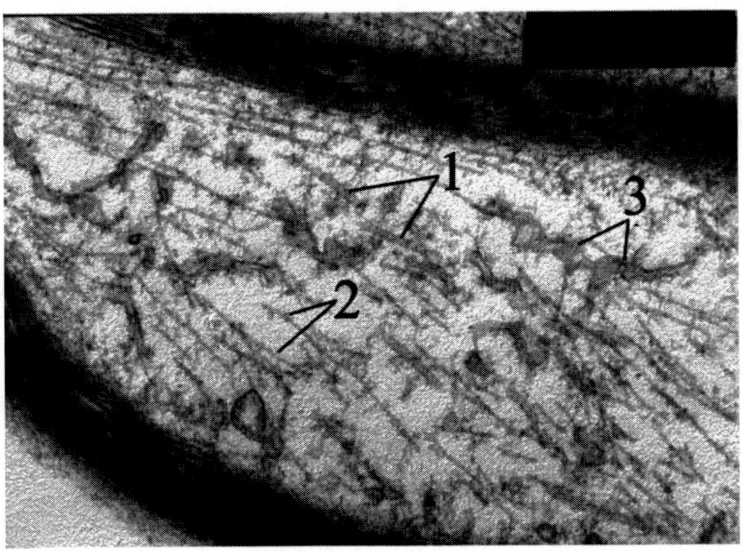

Figure 57. Fragments of the shortened microtubules under the action of colchicine. 1 – microtubules; 2 – neurofilaments; 3 – Endoplasmatic reticulum. Electron microscopy. Mag. 48000.

Under the action of colchicine on neurons, we noted the inhibition of the contraction of nerve fibers for 18 hours: it was both complete and irreversible (Figure 56). At full blockage of retraction, even the most delicate side branches of the initial appendage did not contract. Retraction bulbs did not form, in general. The three-hour exposure to colchicine of the neuromembrane demonstrated reactive mobility retraction of nearly all the nerve fibers.

Under electrophysiological studies (Figure 56 a) the action potential was slightly decreased. The frequency of spontaneous impulse activity was sharply increased (73%). The excitation threshold did not change and the latent period increased by 34%. This indicates that colchicine exerts an excitatory, irritating effect on the neuromembrane, but these changes are small and enable the use of this blocker in experiments on the inhibition of the formation of diastase of transected nerves.

Under the action of colchicine, as has already been noted, the polymers of the axoplasm undergo a restructuring. A decrease in the length of the microtubules is noted (Figure 57). This can change the 'flows' of axoplasm.

Apparently, the tubes have substantially increased the degree of adhesion of single proteins associated with them. Besides this, there is an increase in the degree of binding, and perhaps of the number of related proteins (Figure 58). The proliferation of associated proteins apparently also enables the inhibition of the retraction of fibers.

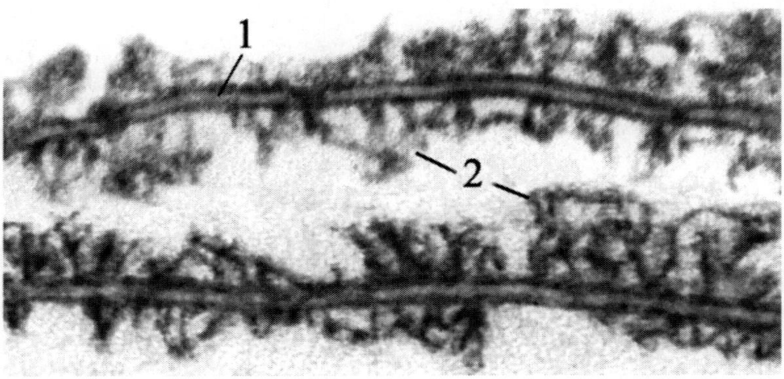

Figure 58. Fragments of microtubules with enhanced adhesion and an increase in associated proteins. 1 – microtubules; 2 – proteins, associated with microtubules under the action of colchicine. Electron microscopy. Mag. 48000.

Major changes occur in the cisterns of the endoplasmic reticulum. The tubular apparatus of the reticulum simultaneously develops and degrades. The concentration of tubules of the reticulum increases and the number of their associated proteins grows (Figure 59).

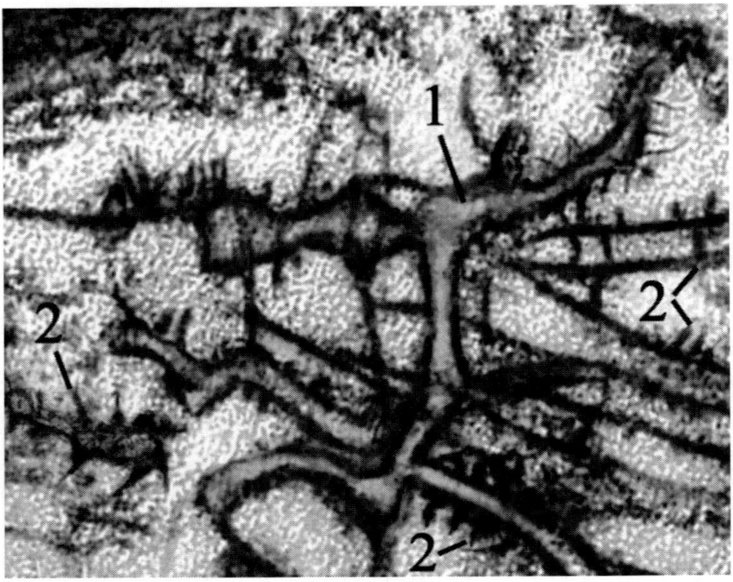

Figure 59. Increase in the density of the distribution of tubules of the endoplasmic reticulum under the action of colchicine. 1 – microtubules of the reticulum; 2 – neurofilaments. Electron microscopy. Mag. 48000.

Concurrently with the intensive growth in the quantity of the tubules there is also a collapse (Figure 60).

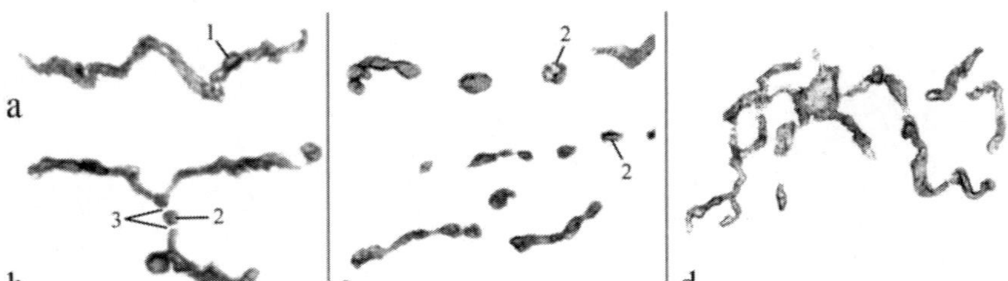

Figure 60. Non-specific feature of the effects of colchicine, fragmenting of the endoplasmic reticulum of the axon. a - varicose deformation of the cistern of the endoplasmic reticulum; b - the isolation of residual cells from the cistern; в - the fragmentation of the tank to separate bodies; d - the division of the reticulum network into separate fragments. 1 - varicosity; 2 - residual bodies; 3 - forming pores between varicose elements. Electron microscopy. Mag. 20000.

Initially, the varicose-like structures form on the tubule. It is then separated as an independent bubble, finally breaking into fragments. A significant portion of the endoplasmic reticulum can simultaneously fragment.

Within three hours of exposure to cytochalasin B the amplitude of a single pulse increased by 9.6%. However, systematically, together with single spikes, double spikes appeared with low amplitudes (Figure 61).

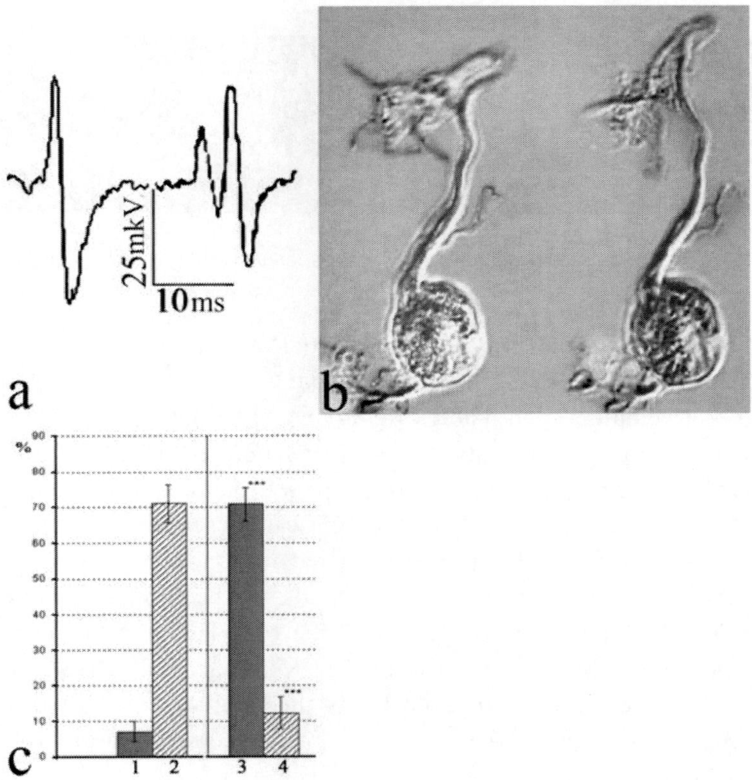

Figure 61. Electrical and morphological effects of the action of cytochalasin B. a – changes in action potential; b – absence of retraction of the nerve fibers; c – percentage content not contracting in the cytochalasin B and contracting in the control group of nerve fibers; non-contracting fibers (1, 3); contracting fibers (2, 4). Note: *** - reliable significant variations from the control at a level of significance r ≤ 0,005.

Then the total value of the amplitude of the spikes was approximately equal to the amplitude of a single impulse. The frequency of the background impulse activity, the threshold for synaptic stimulation and the latent period increased. The growth in the frequency of impulse activity by 87% was probably due to the appearance of double spikes (Figure 61 a). The increase in the threshold and the latency period (26% and 52%, respectively) did not cause any damaging effects on neurolemma because the amplitude of the single spikes remained the same. This indicates the absence of a pathological influence of this blocker of motor activity on the neurites.

Significant changes were recorded amongst the proteins associated with cytochalasin *B* (Figure 62). 70% of axonal contraction was blocked.

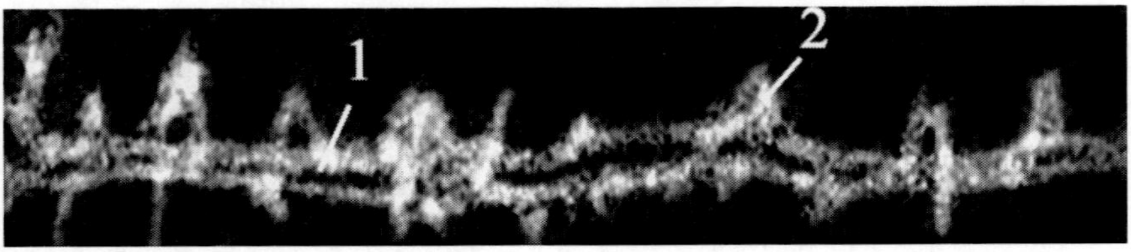

Figure 62. Fascicle of neurofilaments (1) with associated proteins (2) under the influence of cytochalasin *B*. Electron microscopy. Mag. 48000.

In this way, the action of cytochalasin B clearly blocks the movement of a large quantity of isolated fibers. The quantity of non-contracting neurites under observation between 5 and 13 hours increased tenfold (from 7% in the control to 70%). The quantity of axons reducing their contractile activity decreased from 71% to 13%. The latent period of their contraction under a concentration of cytochalasin B of 0,2 mm was preserved for up to 4 hours. Therefore, cytochalasin B can be considered fully applicable in attempts to inhibit the retraction of severed nerves and also in the clinic for the treatment of reducing the function of phagocytosis (Banville, Fallon, McLoughlin et al. 2011) and osteoarthritis (Sauter, Buckwalter, McKinley et al. 2012).

The effect of blebbistatin, a blocker of the contractile activity of myosin, manifested itself in the prevention of 70% of the retraction of the neurite, the same as cytochalasin (Figure 61, 63). The similarity of the effects of cytochalasin B and blebbistatin is apparently due to their overall operation as part of actomyosin. When using blebbistatin in complete darkness spontaneous pulse amplitude increased by 37%, the frequency of the background impulse activity insignificantly decreased by 25%, and the threshold and the latency period of the stimulus increased significantly by 35% and 52%, respectively. Thus, blebbistatin provides a braking, apparently hyperpolarizing influence on the electrophysiological properties of neuromembranes, which cannot be regarded as pathological. During the use of blebbistatin for three hours it was necessary to observe the condition – carrying out the experiment in total darkness.

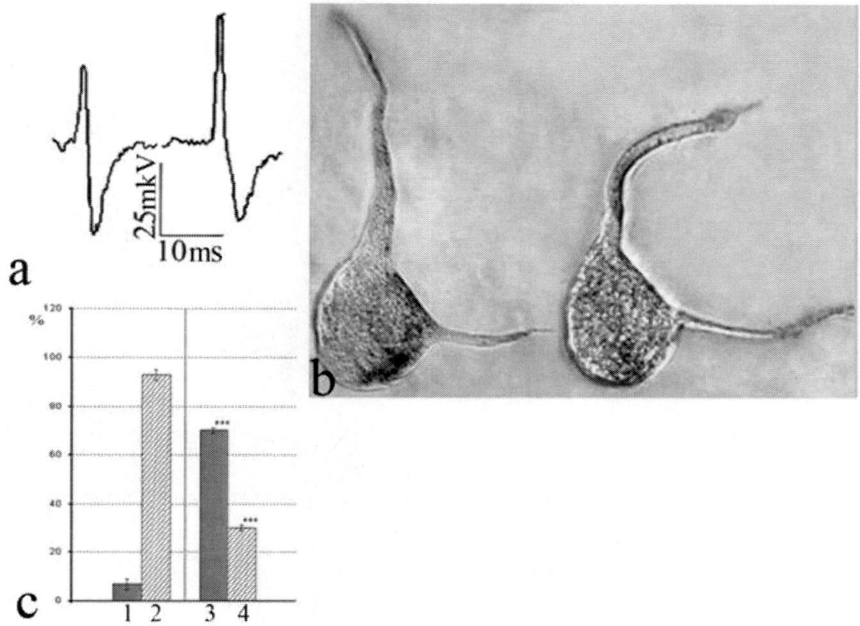

Figure 63. The electrical and morphological effects of the action of blebbistatin. a – increase in the activity potential; b – absence of the retraction of nerve fibers; c – percentage of the content, not contracting in blebbistatin but contracting in the control of nerve fibers; non-contracting fibers (1, 3); contracting fibers (2, 4). Note: *** - reliable significant variations from the control at a level of significance r ≤ 0,005.

Under the influence of the concentration of 50 uM blebbistatin, the electrical properties of neurons significantly increases (Figure 63) the amplitude of the spontaneous spike from (49.5 ± 1.08 mV to 68 mV ± 3.1). The frequency of the background impulse activity decreased insignificantly from 0.24 ± 0.04 pulses/sec from the norm to 0.18 ± 0.04 pulses/sec. The threshold of cells for synaptic stimulation was significantly increased from 70.4 ± 4.9 mV to 95.3 ± 6.4 mV. The latent period of response to an irritating stimulus significantly increased from 22.1 ± 0.7 ms to 45.9 ± 2.9 ms.

The calcium channel blocker nimodipine L-type also has a distinct structural action, blocking the contraction of neurites. Within three hours it also prevented the contraction of 72% of the fibers (Figure 64). Use of the calcium channel blocker of L-type nimodipine in darkness for three hours caused a slightly significant decrease (28%) of the amplitude of spontaneous pulse and frequency of pulse activity (40%).

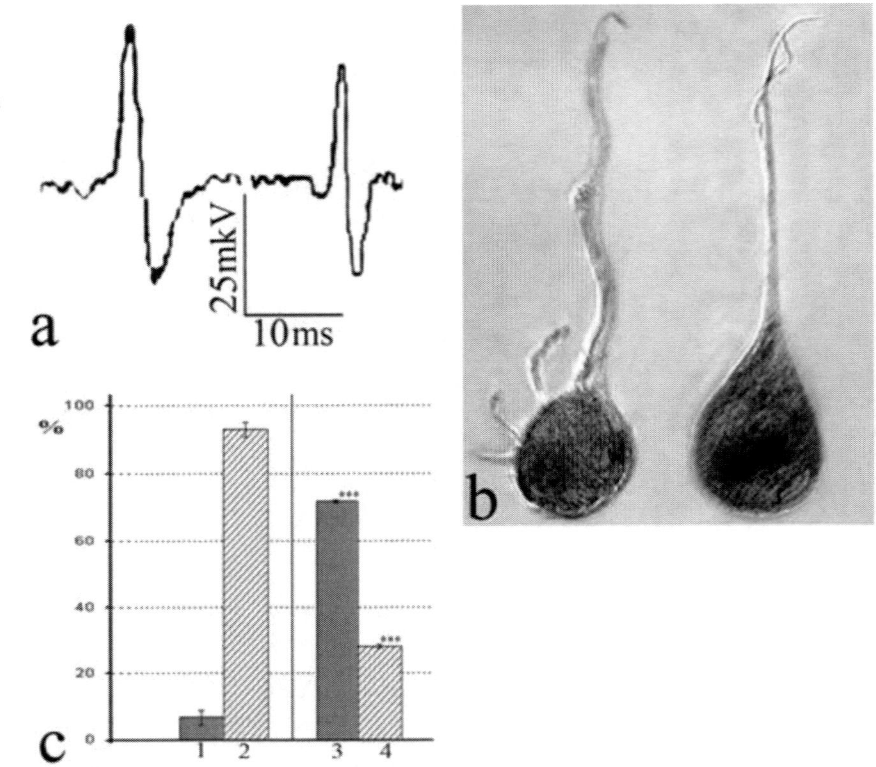

Figure 64. Blocking action of nimodipine on living axon. a – a reduction in the action potential of fiber under the influence of nimodipine; b – intravital microscopy. Phase contrast. Vol. 40 ph, approx 10; c – quantity of non-contracting fibers (1, 3) and contracting (2, 4) neurites in the control of nimodipine. Note: *** - reliable significant variations from the control at a level of significance r ≤ 0.005.

The threshold of synaptic stimulation and the latency response to an irritating stimulus was significantly increased (by 29% and 10%, respectively). Apparently, nimodipine has somewhat of a braking effect on neuromembranes, but it can be used in experiments to block the traumatic retraction. It has already been used in many diseases (Pugh, Berg, 1994; Dispersyn, Nuydens, Borgers et al. 1999).

So, diastasis can really be reduced in the post-operative period using colchicine, cytochalasin B, blebbistatin and nimodipine. They may be tested in experiments on whole animals to counter the expanding diastasis (Table 1).

We have tried to summarize and present the experimental results obtained on single fibers of living vertebrate and invertebrate animals, as the hypothesized process of formation of diastase, scar neuroma and nerve transection in a schematic diagram (Figure 65).

As is known, initially, at the same time as an injury, there is some reduction in the length of the stump of nerve fascicles, which occurs as a result of the mechanical action of the elastic properties of the connective tissue membranes of the nerve. Following this, irreversible contraction of the traumatized neurites occurs.

Table 1. General characteristics of the state of neuromembrane

Indicators of the parameters of the neuromembrane	Nimodipodine before and after	Cytochalasin B before and after	Blebbistatin before and after	Colchicine before and after
Amplitude of the spontaneous spike (uM)	47.5±2.8 34.0±4.5	46.8±0.84 51.8±2.17	43.5±1.08 68.0±3.1	46.9±0.8 37.4±1.3
Frequency of spontaneous impulse activity (pulse/sec)	0.20±0.05 0.12±0.03	0.30±0.05 0.56±0.14	0.24±0.04 0.18±0.04*	0.2±0.04 0.75±0.12
Threshold for synaptic stimulation (мV)	69.3±2.06 89.0±2.36	56.0±3.8 83.0±4.3	70.4±4.9 95.3±6.4	54.3±6.2 42.1±5.2*
Latent period of response (ms)	18.2±1.2 20.1±0.6	24.2±0.7 37.6±1.9	22.1±0.7 45.9±2.9	25.0±0.7 37.0±1.7

* – inaccurate quantities.

Their retraction occurs simultaneously in opposite directions in the proximal and distal segments of nerve fibers, increasing the gap between the severed nerve stumps twofold. This means that the nerves, in principle, can never be firmly sewn or glued since during the operation, the membranes connect, but the axons within them retract. Retraction cannot prevent any tubulation or transplant. As shown in cultures of neurons the regeneration of fibers begins only after retraction of a substantial portion of the axon (Sotnikov, 2012). Judging by experiments on living fibers, the distance to which neurites retract can be significant. Short unmyelinated fibers can be completely retracted into the body of the neuron. Once again confirming the accuracy of the theory of “the bi-directional flow of axoplasm” described by L. Lubin (Lubinska, 1964).

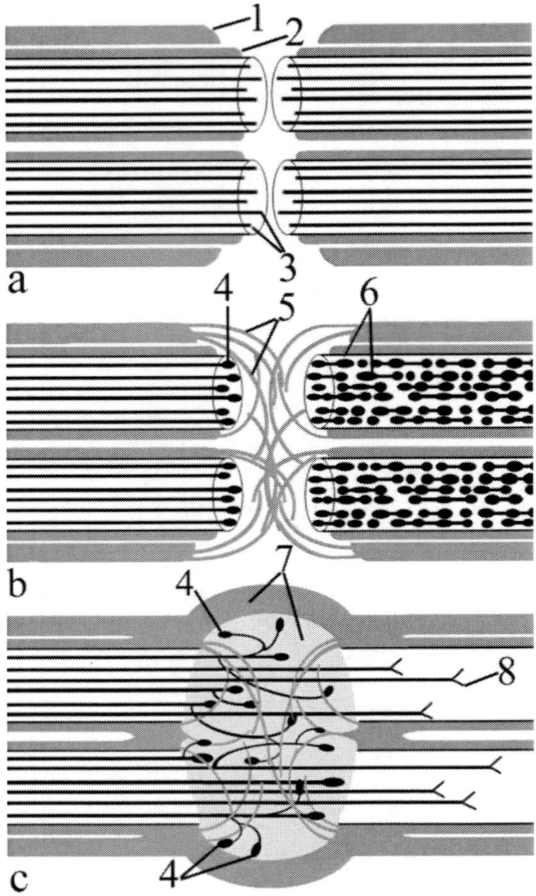

Figure 65. Schematized image of post-operative structural processes in the nerve diastase. a - the initial mechanical, passive reduction in the length of the severed nerve stumps; b - an increase in the width of the diastase by the bi-directional contraction of nerve fibers and the introduction into them of fibrous tissue; c - retraction of nerve fibers during regeneration in the area of the fibrous scar; 1- epineurium; 2 - perineurium; 3 - axons; 4 - retraction bulbs; 5 - fibers of the connective tissue; 6 - degenerating nerve fibers; 7 - the formation of fibrous scar and neuromas in the area of the diastase; 8 - single anterograde regenerating fibers.

This phenomenon explains the bidirectional retraction of transected nerve stumps. An essential feature of the reduction is the appearance at the ends of the transected axons of both stumps, particularly axoplasmic swelling-retraction bulbs.

Thus, it can be argued that the retraction of nerve fibers must be an essential symptom of injured nerves. This data, in our opinion, explains many of the failures in the treatment of transected peripheral nerves. The mechanism of postoperative diastase is apparently due to the contractile, myo-like function of the neuroplasm of axons, causing its shortening. Thus, in summary, the idea of the need of inhibition or blocking of axonal retraction in order to prevent expansion of the gap between the stumps in the treatment of this pathology presents itself. It should be noted that inhibition of natural postoperative retraction of fibers of the spinal cord has been noted previously in a number of experimental studies (Ben-Yaakov, Dagan, Segal-Ruder et al. 2012). Our experiments indicate the possibility of a new approach in the treatment of postoperative nerve transection. Attempts to deal with post-traumatic scars in the nerves and brain had previously made use of numerous inhibitors of collagen formation and cell transplantation (Turgut, Uysal, Pehlivan et al. 2005; Shen, Wang, Gu, 2014), but the role of the retraction of the fibers in the expansion of diastase apparently remains unexplored.

Despite the fact that in principle, the retraction of unmyelinated fibers in tissue culture and in vivo has been shown repeatedly (Wylie, Chantler, 2003; Suter, Miller, 2011), the retrograde contraction of living fibers of the peripheral nervous system, possessing a myelin membrane of complex structure, has evidently been demonstrated for the first time. The molecular mechanism of retraction of injured axons also remains unresearched. Our use of inhibitors of motor activity of the major proteins of the cytoskeletal of axoplasm and calcium channel blockers of L-type, which ensure its retraction, revealed participation in the formation of diastase, the retraction of nerve fibers, nerve damage to the tubulin of neuro-tubules, actin of microfilaments, myosin and calcium. This allowed us to select and try out these blockers of the activity of the cytoskeleton, which are already used in the treatment of various forms of human medicine.

Microelectrode electrophysiological studies have shown that all of these blockers, which can inhibit axonal retraction, have very little effect on the electrical properties of the membrane: the spontaneous spike amplitude, frequency background activity on the threshold of synaptic stimulation and response latency. This data suggests the low toxicity of the inhibitors used in these concentrations in the early and short use of their application. It should be noted that the proposed new approach in the treatment of diastase does not preclude traditional methods of treating this disease, and they do not contradict them. We believe that these results are sufficient to move from isolated studies of drugs to experiments with the cutting and regeneration of nerves in the whole animal.

The experiments have allowed us to make a number of conclusions. Firstly, it has been proven that the nerve fibers after their intersection exhibit myo-like contractile activity. This suggests previously unexplored active mechanisms involved in the formation of diastase in nerves.

Secondly, we have discovered the propensity of the myelinated nerve fibers vertebrates to retract. It has been proven that an absolute sign of contraction of fibers in fixed preparations are retraction bulbs, which are sure to develop on the ends of nerve transection (including human nerves). We have described the bi-directional contraction of axons, indicating the twofold extension of the gap between the severed nerve stumps.

Thirdly, it has been shown that in the retraction mechanism of nerve fibers all major cytoskeletal proteins axoplasm polymers and Ca ions are involved.

Fourthly, we have investigated in experiments that nimodipine, cytochalasin, colchicine and blebbistatin are capable of inhibiting traumatic retraction of nerve fibers, and are not pathological for the electrophysiological properties of the nerve fibers.

We have formulated a practically important theory about the possibility of the use of inhibitors tested in the early stages of the nerve amputation in animal models to inhibit diastase of the stumps in order to prevent operational scar neuromas and accelerate regeneration. It seems that using axon mobility inhibitors can reduce the length of diastasis, thereby reducing the amount of surgical sutures, and the degree of fragmentation of the regenerating growing neurites across the seam.

Chapter 5

WHY DON'T PERIPHERAL MYELINATED FIBERS SWELL UNDER HYPOTENSION?

As is generally known, the cytoplasm of cells placed in a hypotonic solution, one of the oldest environmental factors, is sure to swell. However, in studies of living myelinated peripheral nerve fibers an almost complete lack of swelling was observed. This strange feature of the fiber, which means it does not swell even at 50-30% hypotonic environment for a long time, remains a mystery.

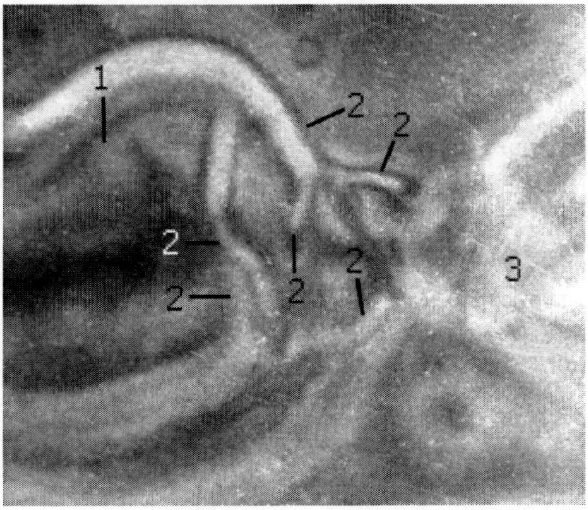

Figure 66. Early changes in the node of Ranvier under stress in Ringer's solution, the splitting of the myelin sheath into large complexes of lamellae. 1 - compact myelin; 2 - complex layered lamellae.

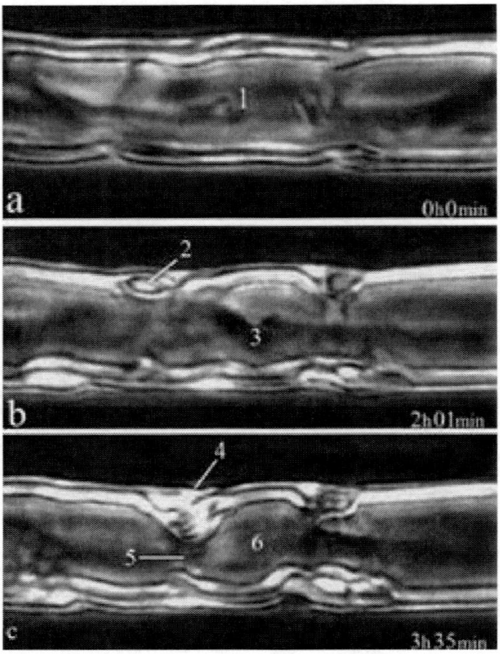

Figure 67. Related reactive changes in Schmidt-Lanterman incisures and axons of myelinated nerve fibers. a-c - dynamics of formation of reactive restructuring of myelin fibers; 1 -wide and smooth cylinder axis; 2 - myelin sheath with subtle incisions; 3 - forming varicosity of the axons; 4 - sharply swollen incision with the delamination of myelin; 5 - local narrowing of swollen axons between incisures; 6 - forming varicosity of the axon. In vivo, serial micro-videophotography. Phase contrast. Vol. 40 Ph, approx. 17.

It is not just that living cells are subordinate to the van't Hoff law, but for completely different reasons. It should be noted that the non-swelling fiber undergoes a series of rearrangements. In particular, their Schmidt-Lanterman incisures increased and their lamellae exfoliated (Figure 66), the size of the paranodal areas of nodes of Ranvier increased (Figure 67) the perikaryon of the Schwann cells (Figure 68), and a strictly local repeated narrowing of the axons (Figure 69). Due to this varicosity forms. The earliest changes in the fiber are intercepting cones and paranodium where the breakdown of myelin first occurs (Figure 69).

Incisions and axons vary simultaneously (Figure 70). However, the outer diameter of the entire fiber does not swell.

We called the entire complex of internal changes the reactive restructuring of fiber. To simplify the problem we will analyze the thick myelinated fibers with a diameter of 7-10 microns. The thin myelinic and non-myelinic fibers require special attention[1]. In experiments with hypotension, the external diameter of the fiber constantly stands at 11,1 ± 0,4 microns, and under hypotension practically remains unchanged, measuring 11.2 ± 0.4 microns. The diameter of the axial cylinder under these conditions in the zone of the swelling incisions decreases to 40.5% (from 7.4 ± 0.3 to 3.0 ± 0.3%).

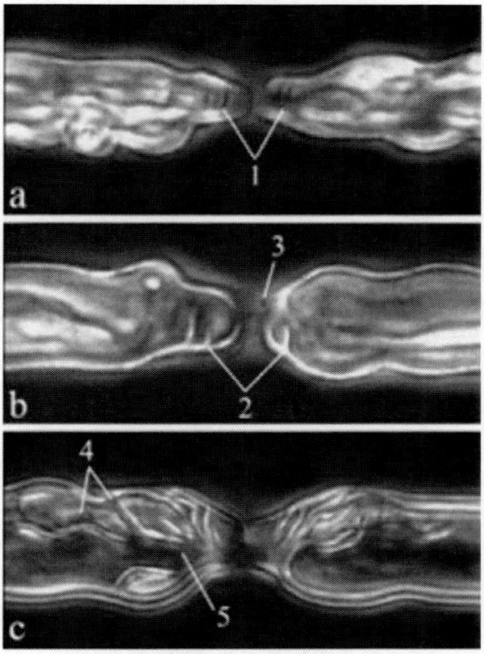

Figure 68. Reactive changes in the nodes of Ranvier. a - transverse striations in the intact cone of constriction, testifying to the original natural separation of myelin; b - an increase in the degree of separation of myelin in the cone interception; a - significant stratification in myelin bulbs of the constriction; 1 - transverse striations in the myelin cones; 2 - group of delaminated lamellar; 3 - the optical effect of the disappearance of myelin in the region of its separation; 4 – delamination of myelin and swelling of paranodium; 5 - formation of the axial cord in the swollen paranodium. Intravital serial microscopy. Phase contrast. Vol. 40 approx. 17.

Like Schmidt-Lanterman incisions, there is separation of myelin in the area of the interception cones (Figure 68). Initially, it is barely noticeable.

[1] Just in case, let us recall that in the thick myelinated fibers in one segment between the heads there are not just present two semi-interceptors (paranodii) in warm-blooded animals there are more than a half dozen Schmidt-Lanterman incisures.

Later the delamination of compact myelin increases and the lamellae intercept complexes become clearly visible. Finally the myelin delamination reaches the paranodium and even the bulb of the interceptor. In the region of the gaps of the interceptor there is a tight rod-like axial cylinder. The paranodium and bulb swell up like the incisions. In the region of the perikaryon of the Schwann cells, the cytoplasm also gradually swells (Figure 69). Upon the swelling of the incisures the axon presents a typical picture of the varicose restructuring of reactive fibers (Figure 70).

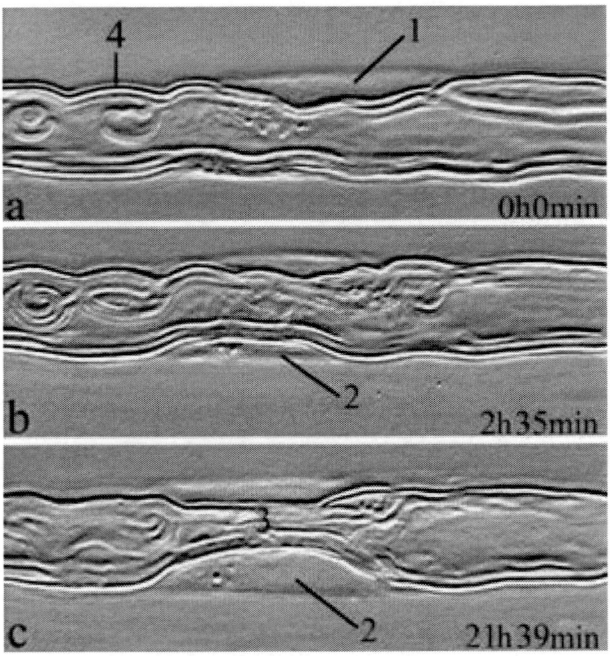

Figure 69. Reactive changes in the perikaryon of Schwann cells. a – intact perikaryon; b – start of the swelling of the perikaryon and the thinning of the axons; c – sharp swelling of the glioplasm on both sides of the axon; d – local thinning of the axial cylinder in the absence of general swelling of the fiber; 1 – perikaryon; 2 – swelling glioplasm; 3 – local swelling of the axial cylinder; 4 – myelin sheath. Serial Intravital video photography. Phase contrast. Emboss effect. Vol. 40 Ph, approx. 10.

It should be emphasized that the incisures and paranodium in three-dimensional space form clearly visible spiral channels upon swelling (Figure 71). If you expand the incisure of myelin, the channels become apparent (Figure 72). They connect the axolemma with the outer layer of the Schwann membrane. However, the modified oval incision and deformed varicose axons in three dimensional space possess irregular cylindrical shapes, consequently we had to carry out additional calculations, representing varicosity and incisures in the form of cylinders. Calculations show that the increased volume of incisures was roughly equal to the volume reduction of the fragments of the cylinder axis. Obviously, there was a mass transfer between these structures. Note that myelin fiber is a symbiosis of Schwann cells and axons of nerve cells. Therefore, these neural-glial relationships should be viewed as a new form of intercellular exchange. It is noteworthy that the total volume of the fiber is maintained throughout. Note two features of this phenomenon. First, it is reversible. All stereometric restructuring of the fiber is quickly restored when the osmotic pressure of the environment is restored (Figure 73).

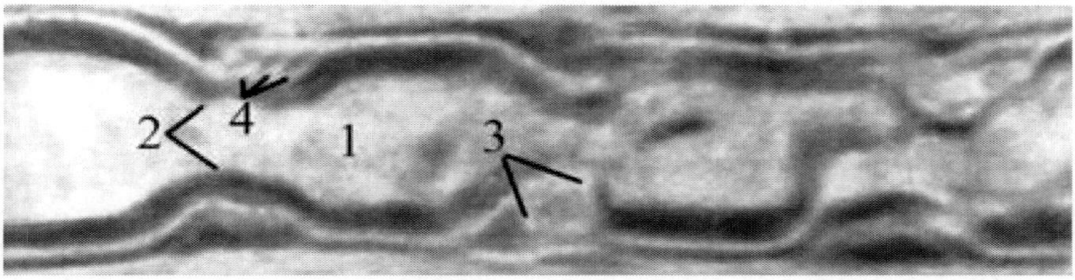

Figure 70. Formation of serial varicosities as a result of local repetitive contraction of the axial cylinder. 1 - varicosity of the axon; 2 - narrowing of the axial cylinder; 3 - swollen incisure; 4 - separating myelin in the swollen incisure. In vivo photography. Phase contrast. Vol. 40Ph, approx. 10.

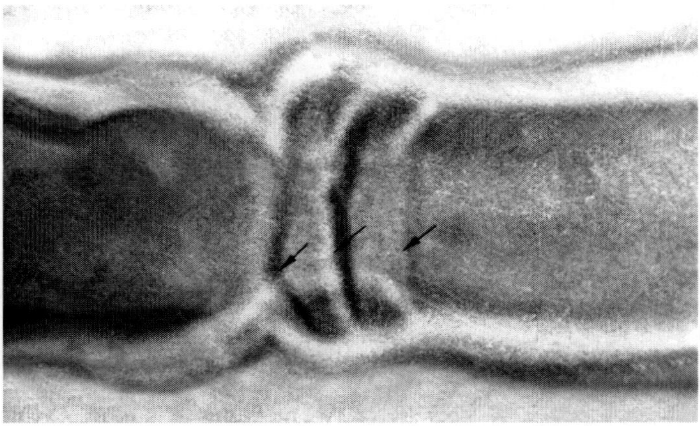

Figure 71. Volumetric cylindrical cytoplasmic canal in the region of incisure (marked by arrows). Intravital microscopy. Phase contrast. Vol. 40Ph, approx. 17.

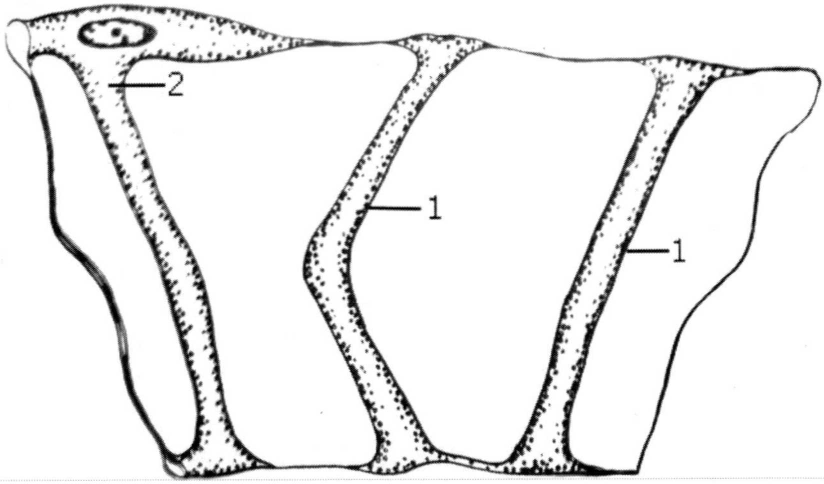

Figure 72. Cylindrical cytoplasmic channels (1) in the region of Schmidt-Lanterman incisions and perikaryon (2). In conditions of the structure of the expanded myelin sheath (diagram).

Secondly, the phenomenon of the simultaneous swelling of incisions and the local narrowing of the axon emerges as a nonspecific reaction to mechanical trauma (stretching), with a long exposure of the preparation in a medium lacking circulation. That is, it is a reversible and nonspecific typical reaction of the fibers to any action.

We can assume that, although the incision swells due to axoplasm faction, some of the water gets into the fiber from the external environment. The quantity of the proportion of this water can be determined by calculating the difference of the swollen volume of the external contour of the fiber and the volume of the swollen incisures (Figure 74). R1-R4 - radii of neural structures of a cylindrical fragment of myelin fiber before and after the reactive restructuring; Va-Ve - volume of neural structures of the cylindrical fragment of myelin fibers before and after reactive restructuring; l - length of the measured cylinder, equal to the length of the swollen incision; the *red color* indicates the possible swelling of the outer contour of the fiber; the myelin sheath in the initial state is colored in *yellow*; the initial state of the size of the axial cylinder is shown in *orange*; the axial structure after changes in the cylinder axis is shown in *brown*; a *gray* parallelogram corresponds to the size of the section of the cylinder equal to the swollen myelin incisures.

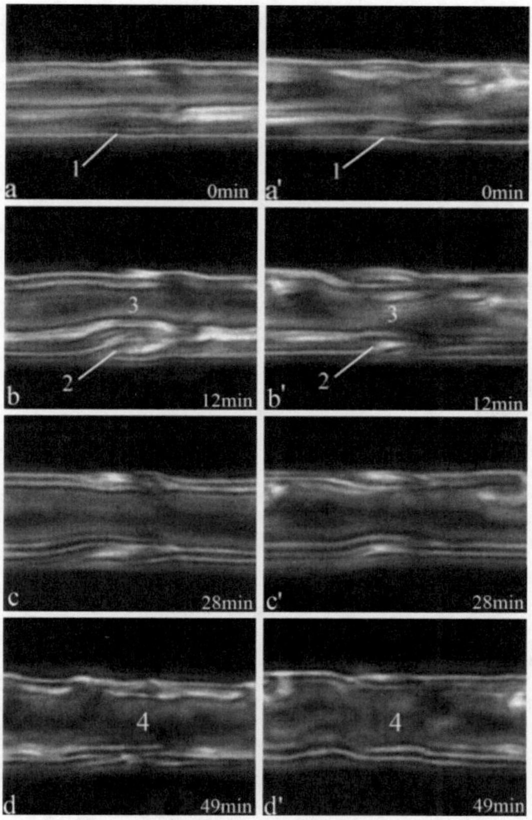

Figure 73. Reactive changes and recovery of different parts of the same segment of internodal myelin fibers after exposure to hypotonic medium. a-d, a'-d'- dynamics of the process; a-a '- the initial state of the two segments of one myelin fiber; b-b '- swelling of the incision and a local narrowing of axial cylinder; b-b ,' c- c' - the restoration of structures of nerve fibers in an isotonic solution. 1 - unaltered incisures; 2 - swelling incisures; 3 - start of the narrowing of the axial cylinder; 4 - the restoration of the axial diameter of the cylinder. Intravital micro videoscopy. Phase contrast. Vol. 40Ph, approx. 16.

R1 - the radius of the fiber after exposure to hypotension; R2 - the radius of the fiber before the effects of hypotension; R3 - the radius of the axial cylinder before the impact of hypotension; R4 - the radius of the tapered axial cylinder after swelling of the myelin incisures (axial strand); Va - the amount of fiber after exposure to hypotension; Vb - the amount of fiber before the effects of hypotension; Vc - amount of possible swelling fibers; Vd - the volume of incisure after swelling; Ve - volume tapered axial cylinder after swelling of the myelin incisures. The volume of the swollen incisures (Vd), as compared with the volume of the fiber, which is increased after exposure to a hypotonic solution (Vc), showed an increase of 5.3 times.

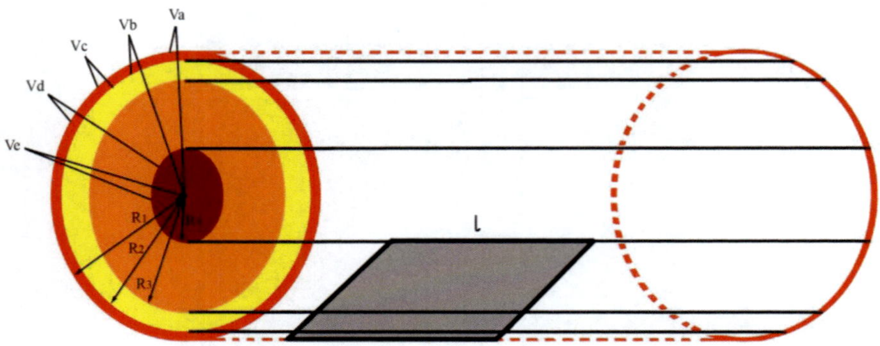

Figure 74. A geometrical model for measuring the volumes of the structures of normal and reactively altered fiber.

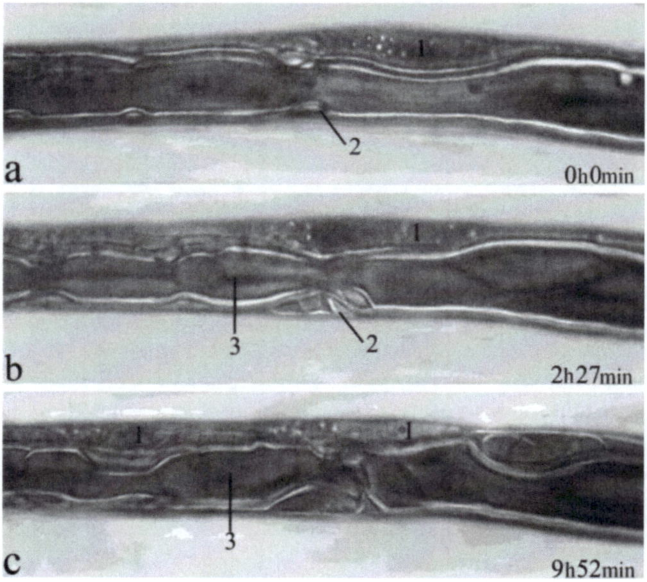

Figure 75. Simultaneous swelling of Schmidt-Lanterman incisures and the perikaryon of Schwann cells in living myelin fibers for 10 hours under anhydrous conditions of perfluorodecalin. 1 - perikaryon of the Schwann cell; 2 - Schmidt-Lanterman incisures; 3 – forming varicosity of the axial cylinder (perfluorodecalin). Intravital serial video photography. Phase contrast. Vol. 40Ph, approx. 17.

In order to provide absolute proof of the use of external water, we considered it necessary to carry out the same experiments in an environment completely devoid of water (water-free environment). For this purpose, we used perfluorodecalin or liquid paraffin[2] as the medium.

It was assumed that in a water-free medium an exchange of external water at its inner fraction would be completely eliminated. The experiment was a success. Despite the absence of water in the environment, Schmidt-Lanterman incisures still didn't swell (Figure 75). When perfluorodecalin is enriched with dissolved oxygen, the process was slowed down and lengthened, stretching out several hours. Thus, you can be sure that the swelling of incisures only happened due to the internal water of the axon.

This presents a new mystery. Where does the axon get the excess water from, and what mechanism causes it to move from the axoplasm to the glioplasm of the incisures and paranodium? When it comes to moving water, primarily the question of its concentration arises. As is well known in colloidal chemistry, osmotic water concentration is not increased by the amount of substance dissolved in it but by the number of particles of the substance in it. (Kaivarainen 1980, Frolov, 1982, Friedrichsberg 1984).

As mentioned in Chapter 1, molecular biological observations have shown previously (Papasozmes, Payne, 1986) that longitudinally oriented protein neurofilaments and microtubules are capable of forming bundles, which reduces the gaps between these structures. Perhaps this is due to the change in the degree of phosphorylation or rearrangement of the conformation of proteins associated with microtubules and neurofilaments. Anyway, they aggregate, reduce the degree of dispersion, i.e., the colloidal osmotic pressure of the fiber reduces.

Apparently, this mechanism of internal redistribution of the water and the reduction of its osmotic pressure is a compensatory mechanism that aligns the osmotic pressure of the axoplasm and the glioplasm with the external environment, and prevents the general external swelling of the fibers, changing its outside diameter.

As shown by electron microscopy, the degree of aggregation can reach considerable intensity, particularly in the area of local restrictions of the axon, turning this area into a real filamentary-tubular strand. So there is an osmotic pressure *gradient* between the axoplasm and glioplasm, which is the engine that triggers the translocation of the neuron-glial movement of water.

---

[2] Perfluorodecalin is a liquid anhydrous substance, enriched with oxygen and water immiscible widely used as blood substitutes in clinics and in urgent cases (Kuznetsov, 2002).

We carried out automatic electron microscopic examination of ultrathin sections of the longitudinal fibers that occur after mechanical trauma in a preparation of single fibers of split nerves from a frog. Scanning these sections showed that the density distribution of filamentous tubular protein structures of axoplasm, which is their dispersibility, was associated with the degree of reduction of the diameter of the structure. The resulting aggregation of these proteins leads to the release of “free” water weakly bound to the protein, which eventually turns up in the area of the incision, the paranodium and perikaryon, that is, where there are clusters of Schwann cell cytoplasm.

As is known, using a drug such as colchicine can increase colloidal osmotic pressure because it is able to dissociate (cleave) proteins of the axoplasm of tubulins into separate fragments and thus increase the pressure of the axoplasm.

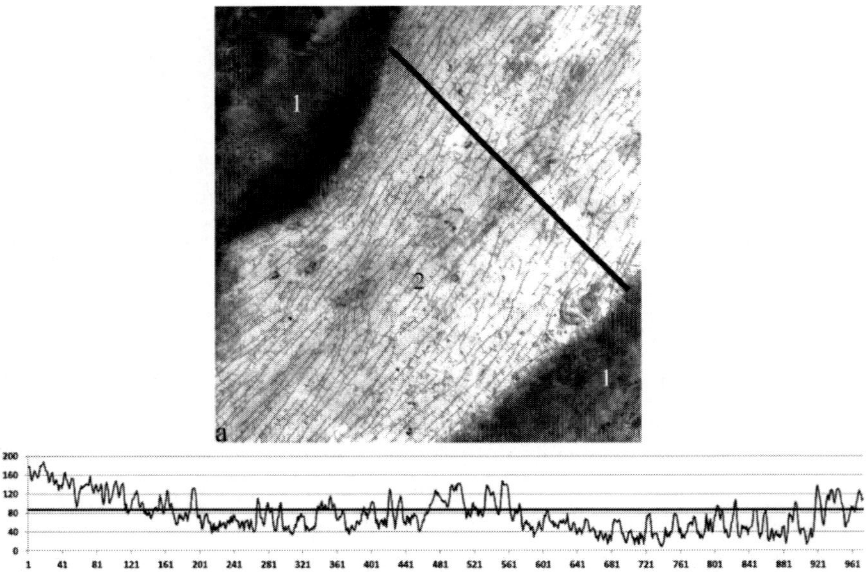

b the density distribution of cytoskeletal structures = 0,040

Figure 76. The density distribution of cytoskeletal structures axoplasm outside myelin incisures. a - the trajectory of scanning the axonal cylinder; b - a graph of the density profile distributions local optical maxima (cytoskeletal structures) along the path of the axon and the numerical expression of this density; 1 - compact myelin; 2 - evenly spaced microfilaments and microtubules; the vertical axis - the optical density in arbitrary units (pixels), which allows to determine the number of optical maxima (number of cytoskeletal structures); on the abscissa - the length of the optical density profile within the scan line, in pixels. Electron microscopy. Mag. 19000.

It should be recalled that the main features of the described reaction is that it has the properties of reversibility and is nonspecific, and colchicine has already been used under many pathological effects (Bahceci, Aluclu, Canorue et al. 2009; Choo, Liu, Liu et al. 2009 ; Kilinc, Gallo, Barbee, 2009). Therefore, in the treatment of diseases of myelinated nerve fibers, and even nerve trunks, there is a fundamentally new unused approach, which is based on intervention in the colloidal chemistry of blood.

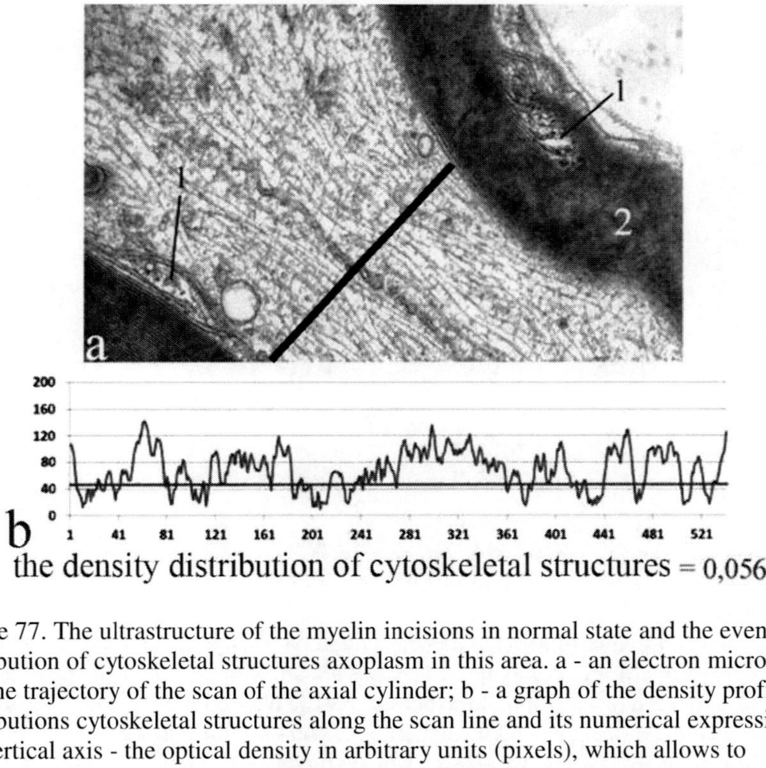

Figure 77. The ultrastructure of the myelin incisions in normal state and the even distribution of cytoskeletal structures axoplasm in this area. a - an electron micrograph and the trajectory of the scan of the axial cylinder; b - a graph of the density profile distributions cytoskeletal structures along the scan line and its numerical expression; the vertical axis - the optical density in arbitrary units (pixels), which allows to determine the number of optical maximums (number of cytoskeletal structures); on the abscissa - the length of the profile of optical density within the scan trajectory in pixels; 1 - normal myelin incisure; 2 - compact myelin. Electron microscopy. Mag. 24000.

It was found that the density of the distribution of structural and tubular optical maximums are 62% greater in injured than in intact fibers. The density of distribution of local optical maxima of cytoskeletal structures the electron diffraction of intact fiber outside the incisure was equal to 0.040 and was calculated as the ratio of the density of local optical maxima to the length of the scan lines (Figure 76). In the normal myelin incisures where islands of non-swollen Schwann cell cytoplasm are clearly visible (Figure 77 a), the density of distribution of these cytoskeletal structures of axoplasm also remained low (0,056) (I phase).

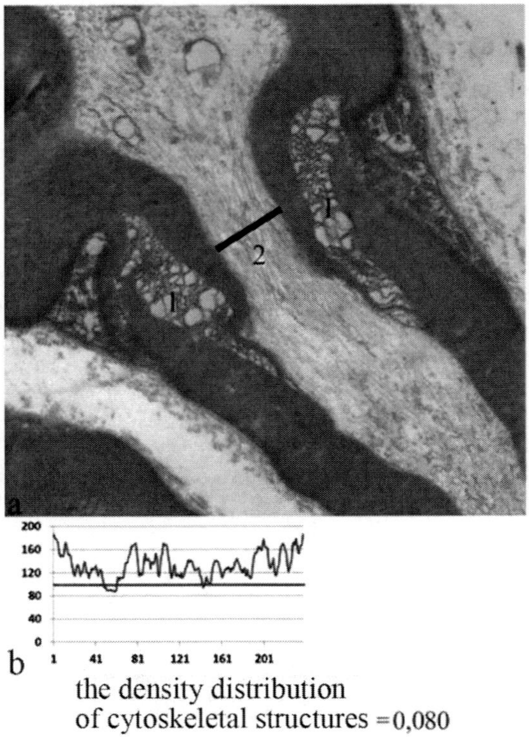

Figure 78. Ultrastructure of swollen incisures (reactive changes of the second phase) and the density distribution of axoplasmic cytoskeletal structures in the area. a - an electron micrograph; b - a graph of the density profile distributions of cytoskeletal structures along the scan line and its numerical expression; the vertical axis - the optical density in arbitrary units (pixels), which allows to determine the number of optical maxima (number of cytoskeletal structures); on the abscissa - the length of the profile of optical density within the scan trajectory in pixels; 1 – swelling incisure with splitting of the complexes of lamellae of myelin; 2 - aggregation of cytoskeletal structures in the area of the tapered axial cylinder. Electron microscopy. Mag. 14000.

Using an electron microscope on lightly damaged constrictions (II phase) observed weak separation of compact myelin into single uniform myelin lamellae (paranodal channel islands) were observed (Figure 78 a). Morphological changes of the swelling of incisures (II phase) were expressed by an increase in their volume accompanied by a local thinning of the axon in the same area. The density distribution of cytoskeletal structures increased to 0.80 units. (Figure 78 b). With a sharp swelling of incisures (Figure 79, a) (III phase), there is a sharp separation of myelin lamellae and a significantly greater thickening of the axial strand (0,107 units).

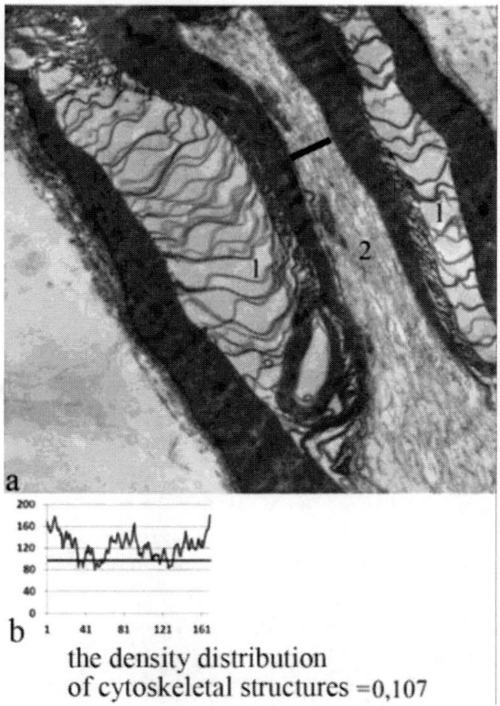

Figure 79. Ultrastructure of swollen incisures (reactive changes of the third phase) and the density distribution of axoplasmic cytoskeletal structures in the area. 1 – sharply pronounced separation of the lamellae of myelin; 2 – forming of the axial strand of dense cytoskeletal structure. Other symbols are the same.

The density distribution of cytoskeletal structures of the axoplasm in intact constrictions reached 0,059 units. (Figure 80a). That is approaching the density distribution of the same incisures. At constrictions of II phase of damage the density distribution of structures of axoplasm increased to 0,084 units. Sudden changes in the injured constriction (III phase) pronounced swelling of the myelin occurs, splitting of its tabular complexes, narrowing the axial cylinder (Figure 79, a) and cytoplasmic swelling of the paranodal channel region. The area occupied by myelin expands, due to the diameter of the cylinder axis.

In this area, the distribution density of cytoskeletal structures of the axoplasm increases to 200% - 250% (Figure 78 b, 79 b), that is to say that filamentous-tubular cytoskeletal bunches form equal in size to the outlet opening of constricted cones in normal state or in the gaps between the swollen incisures.

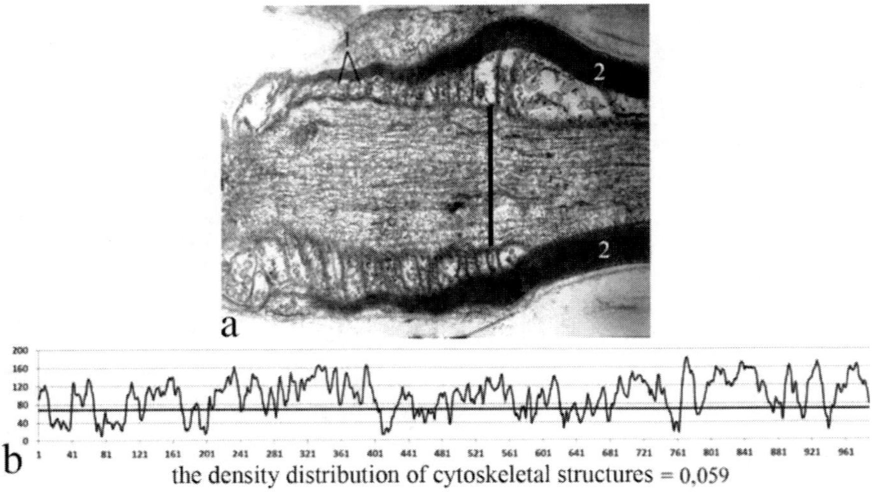

Figure 80. The ultrastructure of the node of Ranvier in normal state and the even distribution of cytoskeletal structures of axoplasm in this area. a - an electron micrograph; b - a graph of the profile of density distribution of cytoskeletal structures along the scan line and its numerical expression; the vertical axis - the optical density in arbitrary units (pixels), which allows to determine the number of optical maxima (number of cytoskeletal structures); on the abscissa - the length of the profile of optical density within the scan trajectory in pixels; 1 - Lamellar separation of myelin in normal state (paranodal loop); 2 - compact myelin. Electron microscopy. Mag. 14000.

In other words in the local areas of swollen incisures and reactively changed constrictions, dense filamentary-shaped tubular aggregates of cytoskeletal protein structures differing sharply from the normal axoplasm with a relatively low concentration of proteins form.

The high density distribution of cytoskeletal structures of axoplasm within the swelling perikaryon of the Schwann cell (0.076, Figure 83 b) compared to the control fiber increases to 0.114 (Figure 84 b). Its increase in comparison to the density of the normal axon outside myelinated structures is also about 200% -275%.

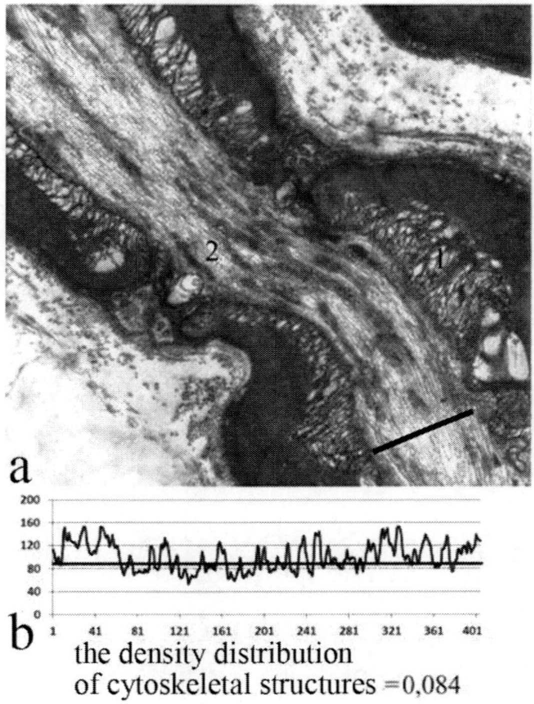

Figure 81. The ultrastructure of the node of Ranvier (reactive changes of the second phase) and the even distribution of cytoskeletal structures of axoplasm in this area. a - an electron micrograph; b - a graph of the profile of density distribution of cytoskeletal structures along the scan line and its numerical expression; the vertical axis - the optical density in arbitrary units (pixels), which allows to determine the number of optical maxima (number of cytoskeletal structures); on the abscissa - the length of the profile of optical density within the scan line in pixels; 1 - significant Lamellar separation; 2 - aggregation of cytoskeletal structures and organelles in the constriction gap and paranodium. Electron microscopy. Mag. 24000.

The same degree of increase in the density distribution of altered cytoskeletal structures in the axoplasm of swollen incisions of the paranodal zone and near the perikaryon of the Schwann cell indicates that the remainder of the dehydrated axon is a single structure – the axonal strand. Thus, the density distribution of cytoskeletal structures of the axon depends entirely on the degree of mechanical damage to the fiber structure (Figure 85).

If our deliberations are true, they can be tested in an experiment using a preparation we've tested – a blocker of the mobility of axons - colchicine (Sotnikov, Vasyagina, Sergeeva, 2013).

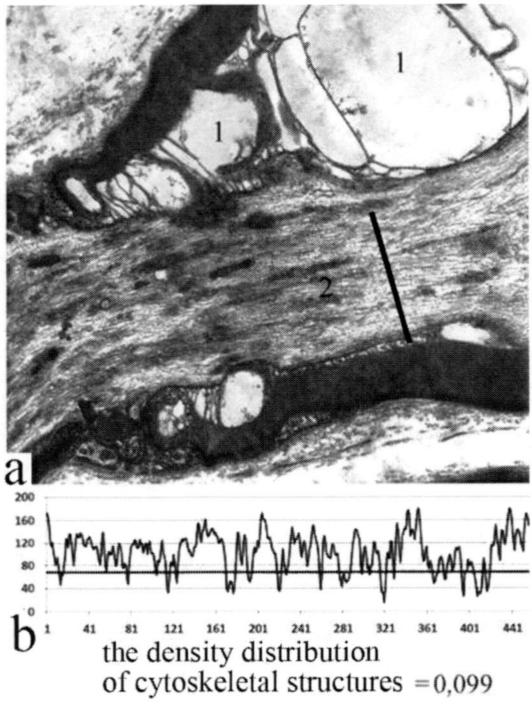

Figure 82. The ultrastructure of the node of Ranvier (reactive changes of the third phase) and the even distribution of cytoskeletal structures of axoplasm in this area. 1 – clearly defined separation of the lamellae of myelin; 2 – increase in the concentration of structures of axoplasm forming an axial strand. Other symbols are the same.

Splitting the protein polymers of tubulin, and thereby increasing the number of molecules and fragments, should raise the degree of dispersion of the solution and thus increase the colloidal concentration of axoplasm. Given that the process of the reactive reconstruction of myelinated fibers begins with reducing the dispersion of axoplasm and a fall in its colloid concentration, swelling of the incisures and paranodium with splitting of myelin in the local constriction of the axon, it is logical to assume that colchicine will inhibit the start of the whole process of the reactive restructuring of myelin fibers.

In order to cause a reaction in the experiment fibers, we used mechanical trauma, which the fibers were subjected to after thorough preparation. We divided them according to the degree of reactive restructuring from normal, not damaged at all by splitting and fibers with significant swelling of the incisure and formation of varicose axons. In the control unmodified and modified incisures appeared (Figure 86) almost equally (48.9% and 50.0%).

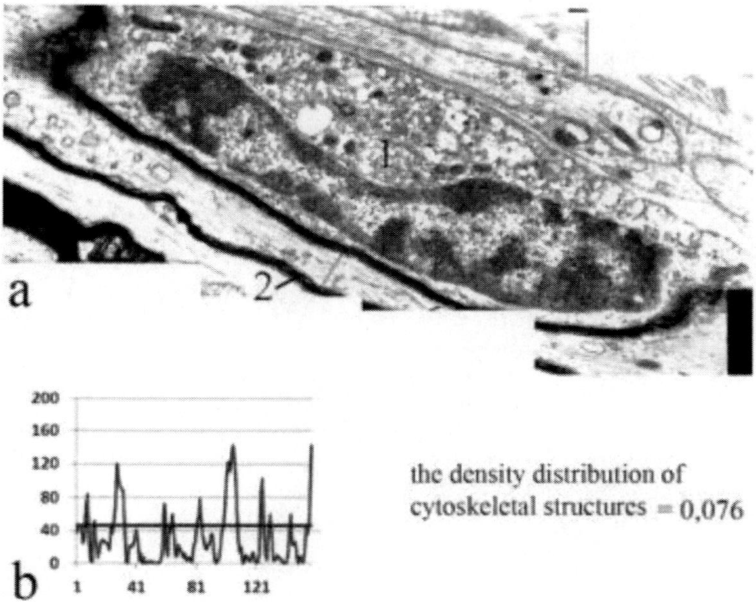

Figure 83. Ultrastructure of the swelling of the perikaryon of the Schwann cell and the distribution of cytoskeletal structures of axoplasm in this area. a - an electron micrograph; b - a graph of the profile of density distributions of cytoskeletal structures along the scan line and its numerical expression; the vertical axis - the optical density in arbitrary units / pixels, which allows to determine the number of optical maxima (number of cytoskeletal structures); on the abscissa - the length of the profile of optical density within the scan line in pixels; 1 - Schwann cytoplasm and nucleus in narrowing myelin fibers; 2 - compact myelin. Electron microscopy. Mag. 8000.

There were hardly any seriously changed incisures (1.1%). All fibers are counted as 100% (Kislov, Chekurova, Kataev, 1999). Most incisures were undamaged (78% in control group - 48.9%) were weakly damaged - 17.1% (in the control - 50%), while there were no dramatically altered incisures. The experimental results showed that the depolimerizer of microtubules, colchicine, clearly increased the dispersion of colloidal solution of axoplasm, reduced the osmotic pressure gradient between the incisures and the axoplasm. Thereby preventing morphological rearrangement of the fibers, that is, reducing the degree of mechanical damage. We checked whether colchicine had caused a decrease in the length of the microtubule. To this end, we measured and compared the average length of microtubules (Figure 87) in the control and under the influence of colchicine, which should have dissociated the microtubules. The average length of intact microtubule fibers equalled 11.2 microns, whereas colchicine reduced their length by 2 times (5.5 mm) (Figure

88). Thus, colchicine really dissociated tubulin and (hence), reduces the possibility of osmotic pressure gradient between the structures of fibers.

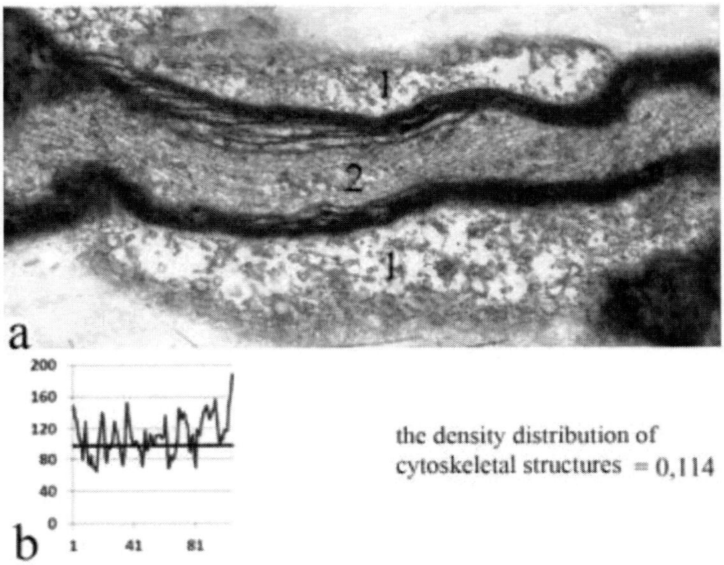

Figure 84. Ultrastructure of the sharp swelling of the perikaryon of the Schwann cell and the distribution of cytoskeletal structures of axoplasm in this area. a - an electron micrograph; b - a graph of the profile of density distributions of cytoskeletal structures along the scan line and its numerical expression; the vertical axis - the optical density in arbitrary units / pixels, which allows to determine the number of optical maxima (number of cytoskeletal structures); on the abscissa - the length of the profile of optical density within the scan line in pixels; 1 – swollen cytoplasm of the Schwann cell from both sides of the stressed fiber; 2 – compact aggregate of cytoskeletal structure in the region of the narrowed axon (axial strand) Electron microscopy. Mag. 10000.

The density distribution of cytoskeletal structures was greater, the greater the degree of reactive changes. Despite the fact that L.A. White, P.W. Baas and S.R. Heidemann (1987) in their research showed uniform distribution of microtubules along the segment of the myelin fibers, our studies show that the density distribution of filamentous-tabular structures in normal fibers still fluctuates, while reversibly swollen incisures, constrictions and perikaryonic Schwann cells increases significantly depending on the reactive changes, i.e., in the area of the tapered portion, between the varicose sections and the opening of the cone constriction.

Increasing the density of the distribution of the protein cytoplasmic structures means an increase in the intensity of their adhesion (aggregation), which is known to be characteristic of the early protein denaturation during damage to the cytoplasm. Protein aggregation is naturally accompanied by the seeping in of the loosely related water of the protein colloid. We believe that this is happening in our experiments. This form of neuron-glia relationships we have identified for the first time, as opposed to other kinds of interactions described in earlier studies (Brunetti, Di Giamberardino, Porcellati et al. 1981; Nicholls, Martin, Wallace et al. 2008; Bay, Butt, 2012).

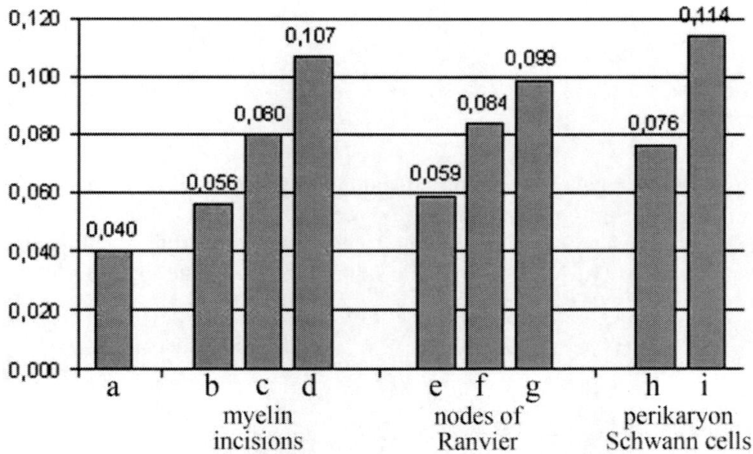

Figure 85. Graphical representation showing changes in the concentration of the cytoskeletal structures of axoplasm (in relative units (pixels)), depending on the degree of reactive changes of the myelinated nerve fibers. a - the distribution density of cytoskeletal structures in normal axoplasm; b - in the Schmidt-Lanterman incisures of normal fiber; c - in the incisures of reactively modified fibers of the second phase; d - in the incisures of the modified fiber of the third degree; e - density of distribution of cytoskeletal structures of axoplasm in the area of the controlled constriction; f – the same in fibers of the second phase of reactive changes of the constriction; g – changes in the concentration of cytoskeletal structures of the constriction in third phase reactive changes; h – density of distribution of cytoskeletal structures in the area of a normal perikaryon of the Schwann cells; i – the same in the area of the reactively changed perikaryon.

Therefore, the mechanism of reactive restructuring of the myelin fiber does not consist is not in the normal swelling of the cytoplasm of Schwann cells, but in the internal translocation of aqueous fraction from axoplasm to glioplasm. Water, filling the swollen structure of the fiber helps to reduce the concentration of its glioplasm and thereby compensates for the swelling of the fibers throughout the experimental environment.

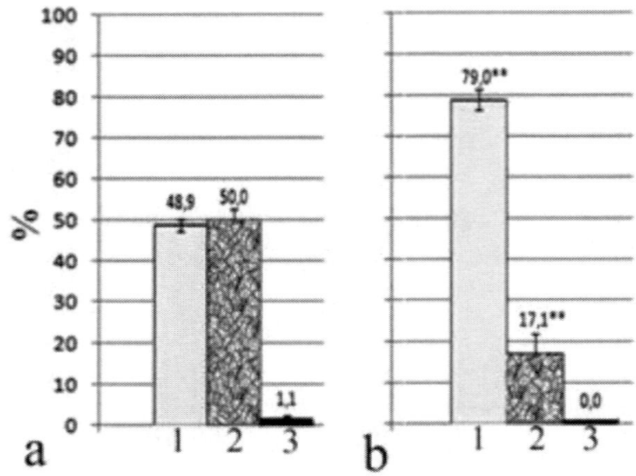

Figure 86. The percentage of unmodified and modified incisures of living myelinated fibers in Ringer's solution, and colchicine. a - control; b - colchicine; 1 - unmodified incisure; 2 - incisures with weak swelling; 3 - incisures with strong swelling and formation of varicosity. Note: ** - reliable significant difference from the control at significance level r ≤ 0,01.

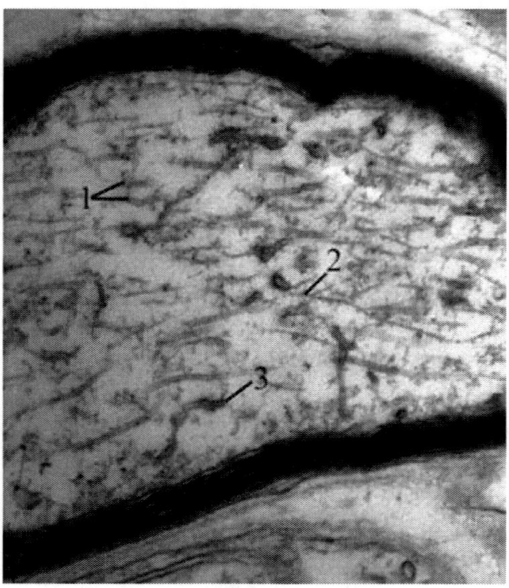

Figure 87. Dissociating microtubules under the influence of colchicine. 1 – microtubules; 2 – neurofilament; 3 – endoplasmic reticulum. Electron microscopy. Mag. 29000.

Thus, the lack of swelling of the myelin fibers in a hypotonic medium can be explained by the fact that the translocation of water from the axial cylinder compensates for and reduces the osmotic pressure in the incisures and the experimental osmotic pressure gradient between the preparation and the external environment.

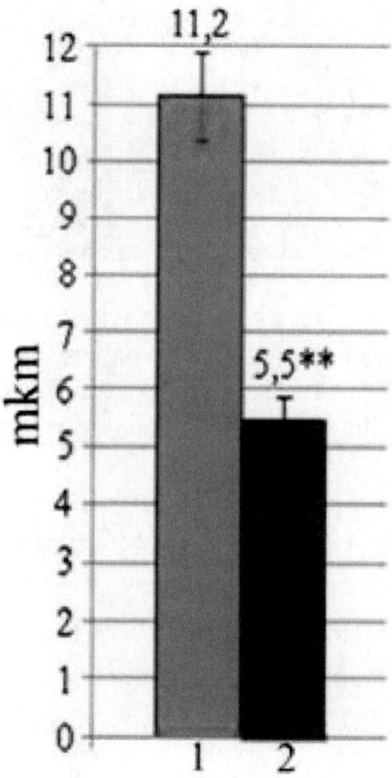

Figure 88. Average length of microtubules (uM) of intact fiber before (1) and after (2) the influence of colchicine. Note: ** - reliably significant variations from the control at a level of significance r ≤ 0,01.

The non-specific mechanism of the restructuring of myelin fibers described above allows the re-evaluation of a range anatomico-pathological changes. In particular, it is possible to analyze the primary kinetics of demyelination, often occurring in many disease processes. Swelling of incisures - is just the beginning of pathology. From the ontogeny of the myelin sheath a circular protein-lipid lamellar layering of the cell membrane forms, followed by their merger into intracellular membrane contacts.

In a number of areas there remain islands of glioplasm, which repeat cycles of forming layers of myelin, forming a single channel (Figure 89). In optical sections it is an incisure with multiple consecutive chambers, with layers of protein, which swell because of the free water from the axon. During this process the protein layers separate a layer at a time or, more commonly, in groups, forming a wonderful picture of symmetric successive chambers.

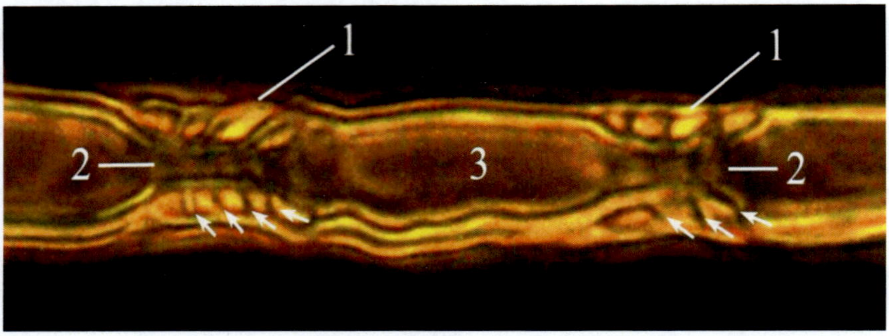

Figure 89. The structure of reactively changed myelin incisures before the process of demyelination. 1 - myelin incisures; 2 - locally narrowed axial cylinder (axial strand); 3 - varicosity of axial cylinder of thick fibers; Arrow - separation of compact myelin into large complexes of lamellae. Intravital microscopy. Phase contrast. Vol. 40Ph, approx. 17.

This process does not stop but continues to spread along the length of the fiber, which loses its geometric orderliness. The disintegrating myelin layers randomly attach to the outside or the inside of the remaining myelin forming a space that some neurohistologers historically used to call Mauthner sheaths (Mauthner, 1868) (Figure 90). The beginning of the formation of the Mauthner area is a result of groups of filamentous, tubular material which is peeled from the compact myelin (Figure 90).

Sometimes, however, the peeling of axonal fascicles from the neurofilament occurs outside the incisures and the destruction of the membrane of the axon (continuing the formation of Mauthner space) (Figure 91).

The pictures depicted by the authors have not been lost but have become part of the factual body of science. Of course, the authors might be mistaken in their interpretation, particularly in the functional significance of structures. However, no matter how many years have passed, the diagrams themselves remain, and they cannot be ignored. They should be interpreted in a new light. The same thing happened with some morphological structures of nerve fibers. They are forgotten and are not referred to. They mysteriously disappeared from the literature. But it is still necessary to uncover this secret of morphology.

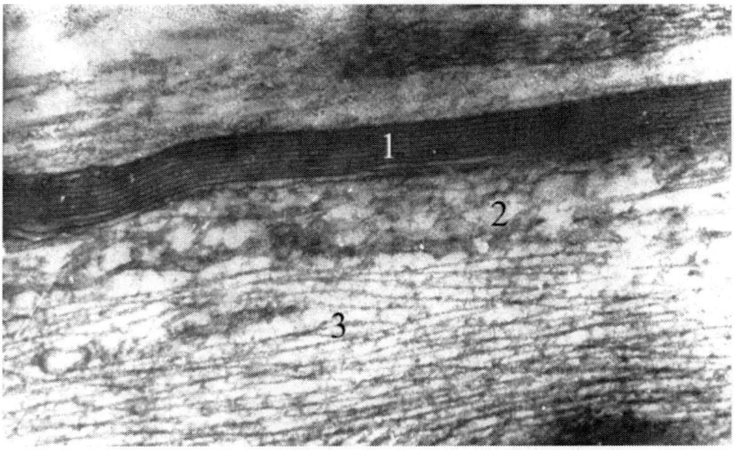

Figure 90. The formation of Mauthner space is caused either by peeling of the filamentous tubular skeleton fascicle from the compact myelin (1), or by the destruction of myelin layered incisures (2).

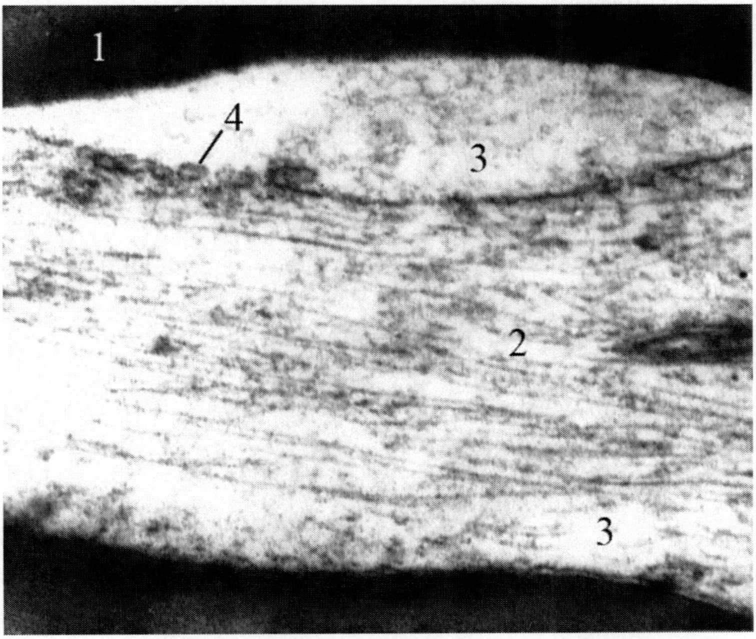

Figure 91. The initial period of detachment of the tubular filamentary strand of the myelin sheath. 1 - myelin; 2 - axial filamentary fascicle; 3 - forming of Mauthner space; 4 - break up of the axolemma membrane.

A cross section of Mauthner space is shown in Figure 92.

Many authors believe that the ancient works of previous authors do not merit further study. I think that in morphology this is not the case.

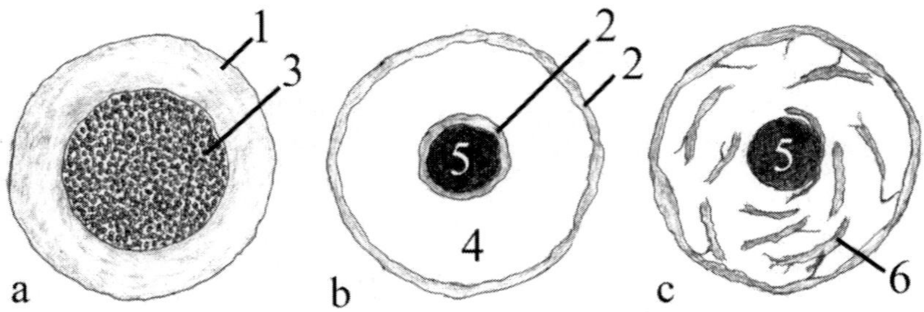

Figure 92. Dynamics of the demyelination of nerve fibers (schematic representation, cross-sections). a - an intact myelin fiber; b - fiber with dramatically swollen Schmidt-Lanterman incisure and sharp narrowing of axons; b - formation of Mauthner space as a result of the destruction of the lamellae of the myelin sheath (demyelination). 1 - myelin sheath; 2 - myelin sheath, separation of the swollen incisure; 3 - axis cylinder; 4 - swollen Schmidt-Lanterman incisure; 5 - aggregate of filamentary tubular proteins (axial strand); 6 - damaged fragments of the myelin sheath during its demyelination (Mauthner space).

The presence of the space between the axon and myelin sheath layers was noted by F. Leydig (1857) in his "Histology of people and animals." Then this space was described by other authors (De Renui, 1929 a, b; Cowdry, 1952; Totibadze, 1958; Voronin Kuparadze, Totibadze, 1960). However, this structure is best known as the Mauthner space, i.e., the space between the axon and myelin sheath. The explanation for this structure has yet to be given, and now in the era of electron microscopy it is largely forgotten. So the formerly mysterious and now forgotten Mauthner "sheath" becomes an understandable pathological structure with its own pathogenesis. Thus this incredible structure, practically absent in all modern textbooks, has found its place. Our persistence was not in vain.

Another fantastic story begs to be told here. It concerns the strangeness of the axon (axial cylinder) of myelinated fibers. Each neurohistologer knows that myelinated fibers axons are 3-4 times thicker than the layer of the myelin sheath. It has successfully and repeatedly been shown on osmified electron microscopic preparations, and the absence of the Mauthner sheath became an undisputed fact (Figure 93 a). But the question arises, where did those thin axons go (Figure 93 b), which are 3-4 times thinner than the myelin sheath on which the whole of light-optical neuromorphology is based? They were there initially. And now, often in textbooks, books and atlases on neurohistology these excellent axons, impregnated with silver salts decorate many foliants (Figure 94). And what about the idols of neurology Kamilo Golgi and Ramon J. Cajal, who created classic neurohistology?

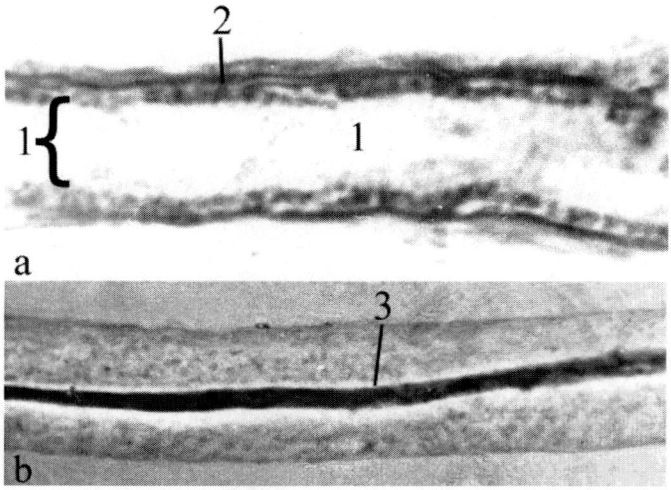

Figure 93. Differences in the axial cylinder upon dyeing of the fiber using various methods. a – osmium oxide staining; b - impregnation with silver nitrate.1 – axial cylinder; 2 – myelin sheath; 3 – axial strand. Vol. 40, approx. 10.

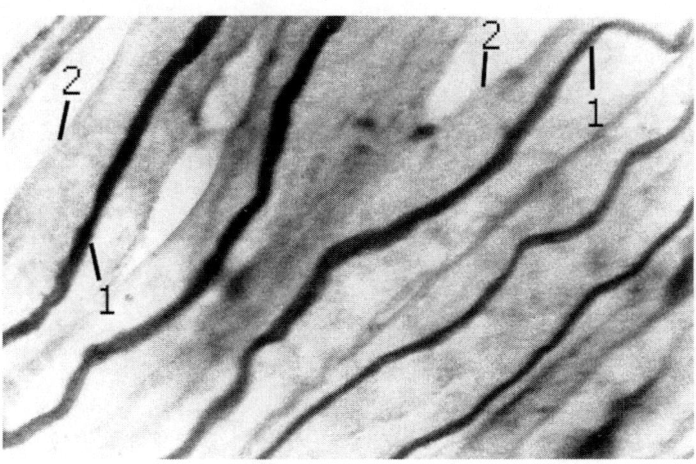

Figure 94. Mass erroneous interpretation of cytoskeletal axial strands (1) as an axon and myelin sheaths profiles (2) (a typical device for textbooks). Golgi technique. The sciatic nerve of the frog. Vol. 40, approx. 10.

Myelinated fibers are uniquely given to analysis, but in fact isn't almost all of neurology premised on preparations impregnated with silver? What is the essence of another neuromorphology puzzle? The morphometry of the openings of retraction cones shows that the thick myelinated fibers have "axons" narrowed (3-4 times) and have the same dimensions as the openings of their own cones, as shown in Figure 95.

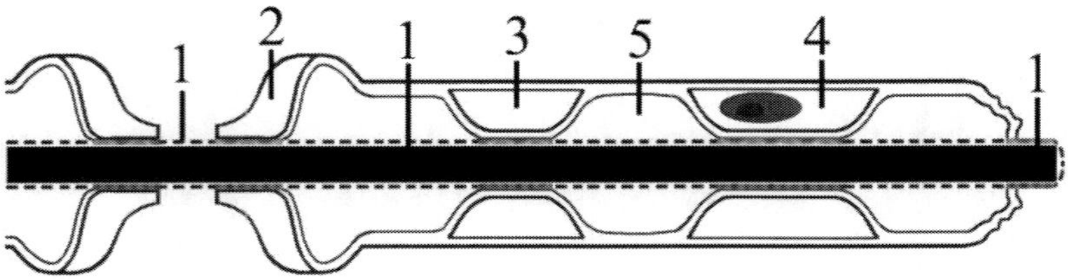

Figure 95. Formation of axial strand within the axial cylinder upon the reactive restructuring of myelin fiber structures. 1 - aggregated axial cord in the area of the gap of the node of Ranvier; 2 - perinodal channels filled with Schwann cytoplasm; 3 - swollen Schmidt-Lanterman incisures filled with Schwann cell cytoplasm; 4 - perikaryon of Schwann cells; 5 - varicosity with the remaining axoplasm.

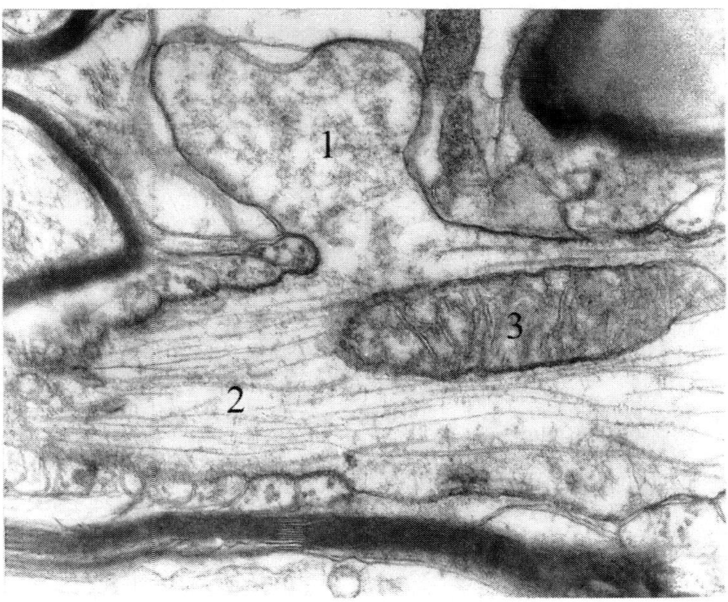

Figure 96. The axial strand in the area of the node of Ranvier upon the invagination of axoplasm. The mechanical stability of the cytoskeletal fascicle in contact with the axoplasm in the area of the node of Ranvier. 1 – invagination of axoplasm; 2 – neurofilamentous fascicles; 3 - mitochondria.

The formation of the axial strand is also clear, but why does it appear? If we go back to the light optical fibers of living, unstained preparations, you will find another oddity. It turns out that the diameter of the axon in the constriction is exactly equal to the diameter of all the bridges between the varicosity formed upon the swelling of the incisures and Schwann cells. In the case of the protrusion of the axoplasm on the sides of the node of Ranvier the cytoskeletal strand remains smooth (Figure 96) and doesn't detour from its path, formed by the cytoskeleton.

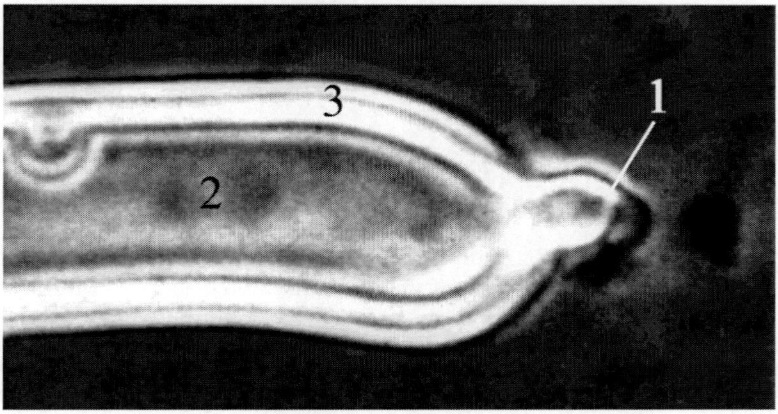

Figure 97. The terminal of the axial strand, extending from the axoplasm after the rupture of the fiber. 1 – axial strand; 2 – axoplasm; 3 – myelin sheath; Intravital microscopy. Phase contrast. Vol. 40Ph, approx. 17.

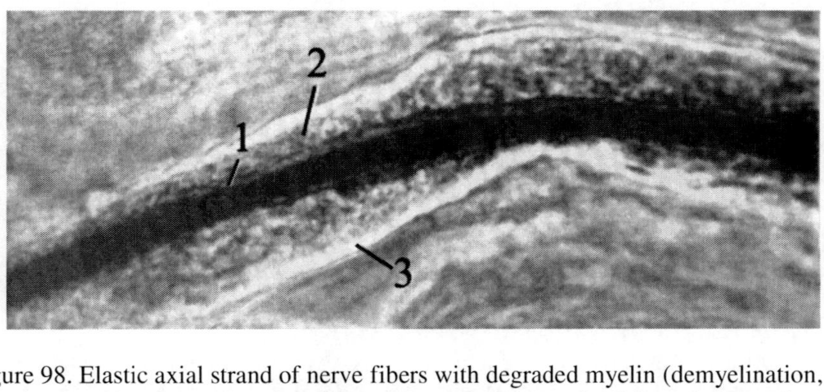

Figure 98. Elastic axial strand of nerve fibers with degraded myelin (demyelination, Mauthner space). 1 - aggregation, filamentous-tubular strand, of the polymers of axoplasm protein; 2 - traces of destruction of the myelin sheath (the second variation of the Mauthner space); 3 - the remaining outer layer of the myelin sheath. Bielschowsky-Gross technique. Vol. 40, approx. 17.

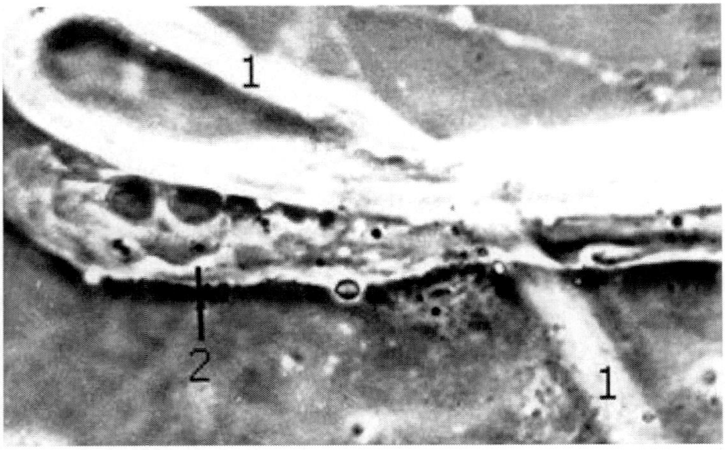

Figure 99. The axial strand (1) demonstrates its mechanical durability and elasticity outside the myelin sheath (2).

Upon the reaction, the swelling of the incisures is restricted, as it were, by the level strand, which compresses the incisure. When the swelling incisures extend, this flat structure also lengthens (Figure 96). In the area of the severed fiber and its retraction this structure contracts more slowly than the entire axon, and therefore acts as a real structure, which we call the axial strand (Figure 97). It has also been noted by A.I. Babukhin (2010). The described structure is a solid elastic strand that is easily bent by turning the non-fixed fiber with degraded myelin (Figure 98) and painted with silver nitrate or methylene blue.

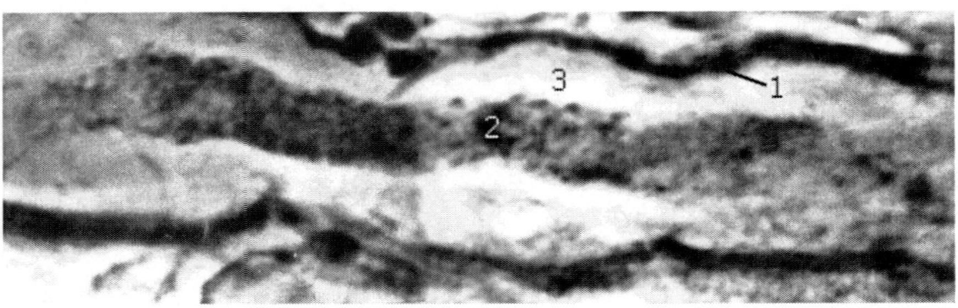

Figure 100. The Mauthner space of non-fixed fibers (demyelination of the fiber). 1 - remainder of the outer layer of the myelin sheath; 2 - aggregating axial cord; 3 - Mauthner space. Phase contrast. Vol. 40, approx. 10.

In rare cases, the axial strand may even be removed from the shell (Figure 99). The swollen myelin sheaths can combine, extend and propagate along the fiber, forming a Mauthner space of considerable length (Figure 100).

Comparative measurements of living, fixed and impregnated fibers clearly demonstrate that axons of fresh fibers are much larger than fixed axons and are exactly equal to impregnated tubular filamentary "silver" strands and cones of the nodes of Ranvier of living fiber "compressed" between swollen incisures.

In our opinion, this is an interesting puzzle for normal neurohistology. The clear formation well-known to all histologists is suddenly not the axon, as many are accustomed to thinking, but the fragment of the narrow, filamentous tubular cytoskeletal strand tightly posthumously aggregated and specifically absorbing silver salt. At the same time, this is a step forward in neuropathology. It becomes clear why in many disease states, the axon becomes so thin. Indeed it is because it is not the axon, but the aggregated axial strand, which always forms posthumously in the known fixative, formalin. It is mechanically the strongest and most stable structure in pathology. The successes of molecular biology are undeniable and very helpful in our time, but it seems to me that classic neuromorphology is not yet complete. Not all its secrets have been disclosed yet, which leads me to my conclusion.

DISCUSSION AND CONCLUSION

The work presented here is an attempt to investigate new properties of axoplasm using living denuded axon preparation of molluscs with a preserved neuron body. The amazing elasticity of the axoplasm, its ability to lengthen over long distances, the presence of her initial retractile propensity and the ability to intensely aggregate under adverse conditions have been described.

Analyses of the protein polymers of axoplasm and experiments with transected preparations have shown that axoplasm is not a liquid and does not flow from the incisions. It cannot flow and retracts, intensely invaginating into the body of the neuron. Experiments with a change in the axon adhesion area revealed the possibility of adjusting the direction of transport of the axoplasm. Similar to the cytoplasm muscle, it retracts with its whole mass including the entire cytoskeleton as a single entity simultaneously in opposite directions. This is the first time the original idea of the Lillian Lubinska of bi-directional contraction has been confirmed on living axoplasm, despite the inaccuracy of the experimenter about the fluid plasma cluster at the ends of the axon. In experiments with the direction of axoplasm, the ability to regulate it by changing the place of adhesion was identified. In these experiments, a new form of axoplasm contraction was discovered - isometric retraction.

In tissue culture, it was possible to once again confirm the repeated retraction of axoplasm simultaneously contracting in opposite directions. A sharp thinning down to sub-membrane dimensions, and elongation of axons was shown, allowing the production of invisible fibers and observation of the physiological processes in which they participate.

Weiss, followed by Lubinsky, tried to explain this phenomenon as congealing of axoplasm. But intravital video microscopy experiments proved that the final nerve ellipsoids are retraction bulbs. That is, the retrograde and anterograde contraction of axoplasm of the end sections of neurites.

Thus, the experiments carried out on living axoplasm and first demonstrating its unique properties allow a series of seemingly impossible processes to be solved, secrets of axoplasm that, hopefully, will be included in a list of major non-electric functions of axoplasm.

The interaction of the cytoskeletal proteins of organelles of axoplasm also turned out to be little studied. It seems that all the known pathological neural processes begin with increasing the degree of adhesion of cytoskeletal filamentous - tubular structures that use associative proteins to form, at first, fascicles and then dense axial strands.

This work was carried out over the course of many years, on the preserved form of the axon with new preparations of denuded axoplasm with the help of graduate students and many staff members of the laboratory of functional morphology and physiology of the neuron. It is my pleasant duty to thank the editors of «Axons» Hiro Yamomoto and Aki Ochiro, our laboratory staff N.Y. Fomina, E.A. Gendina, T.N. Kokurina, T.V. Krasnova, A.A. Laktionova, S.S. Sergeeva, M.V. Smirnova, G.I. Titova and the assistants from friendly laboratories L.I. Archakova, S.A. Novakovskaya, S.D. Solnushkina and V.N. Chikhmana. My deep gratitude goes to the reviewers and the editorial board of the journals "Morphology," "Bulletin of Specialist Biology and Medicine," "Reports of AN," "Successes in Physiological Science" and the publishing houses "Nova" and "Intech." The author expresses his deep gratitude to the Russian Foundation for Basic Research for financial support for long-term studies on the physiology and morphology of axons and the publication of this book.

BIBLIOGRAPHY

Al-Jahdari WS, Saito S, Nakano T, et al. Propofol induces growth cone collapse and neurite retractions in chick explant culture. *Canad. J. Anaesthesia* 2006; 53(11): 1078-85.

Alves MM, Burzynski G, Delalande JM, et al. KBP interacts with SCG10, linking Goldberg-Shprintzen syndrome to microtubule dynamics and neuronal differentiation. *Hum. Mol. Genet.* 2010; 19(18): 3642-51.

Arendt T, Stieler J, Strykstra AM, et al. Reversible paired helical filament-like phosphorylation of tau is an adaptive process associated with neuronal plasticity in hibernating animals. *Neurosci.* 2003; 23(18): 6972-81.

Artiaco S, Tos P, Conforti LG, et al. Termino-lateral nerve suture in lesions of the digital nerves: clinical experience and literature review. *Hand Surg. Eur.* 2010; 35(2): 109-14.

Baas PW, Ahmad FJ. Force generation by cytoskeletal motor proteins as a regulator of axonal elongation and retraction. *Trends Cell Biol.* 2001; 11(6): 244-9.

Baas PW, Buster DW. Slow axonal transport and the genesis of neuronal morphology. *Neurobiol.* 2004; 58(1): 3-17.

Baas PW, Mozgova OI. A novel role for retrograde transport of microtubules in the axon. *Cytoskeleton (Hoboken)* 2012; 69(7): 416-25.

Baas PW, Nadar VC, Myers KA. Axonal transport of microtubules: the long and short of it. *Traffic* 2006; 7(5): 490-8.

Baas PW, Yu W. Creative destruction of the microtubule array. *Cell Cycle* 2012; 11(13): 2420-1.

Babukhin AI. Histology in the lectures of Professor Babukhin (1872/3). Moscow: "Retinoids" 2010; pp. 199.

Bahceci S, Aluclu MU, Canoruc N, et al. Ultrastructural evaluation of the effects of cinnamon on the nervus ischiadicus in diabetic rats. *Neurosciences* 2009; 14(4): 338-42.

Banville N, Fallon J, McLoughlin K, et al. Disruption of haemocyte function by exposure to cytochalasin b or nocodazole increases the susceptibility of Galleria mellonella larvae to infection. *Microbes Infect* 2011; 13(14-15): 1191-8.

Bay V, Butt AM. Relationship between glial potassium regulation and axon excitability: a role for glial Kir4.1 channels. *Glia* 2012; 60(4): 651-60.

Ben-Yaakov K, Dagan SY, Segal-Ruder Y, et al. Axonal transcription factors signal retrogradely in lesioned peripheral nerve. *EMBO J.* 2012; 31(6): 1350-63.

Bessonov LA. Theoretical Foundations of Electrical Engineering. Electrical circuits. 11th edition. Moscow: Gardariki, 2007; pp. 701.

Billuart P, Winter CG, Maresh A, et al. Regulating axon branch stability: the role of p190 RhoGAP in repressing a retraction signaling pathway. *Cell* 2001; 107(2): 195-207.

Bird MM, Lieberman AR. Microtubule fascicles in the stem processes of cultured sensory ganglion cells. *Cell Tissue Res* 1976; 169(1): 41-7.

Bray D. Mechanical tension produced by nerve cells in tissue culture. *J. Cell Sci.* 1979; 37: 391-410.

Bray D. Axonal growth in response to experimentally applied mechanical tension. *Dev. Biol.* 1984; 102(2): 379-89.

Bray D, Bunge MB. Serial analysis of microtubules in cultured rat sensory axons. *Neurocytol.* 1981; 10(4): 589-605.

Brown JA, Wysolmerski RB, Bridgman PC. Dorsal root ganglion neurons react to semaphorin 3A application through a biphasic response that reguires myosin II isoforms. *Mol. Biol. Cell* 2009; 20(4): 1167-79.

Brown SM, Henning S, Wellman CL. Mild, short-term stress alters dendritic morphology in rat medial prefrontal cortex. *Cereb. Cortex* 2005; 15(11): 1714-22.

Brunetti M, Di Giamberardino L, Porcellati G, et al. Contribution of axonal transport to the renewal of myelin phospholipids in peripheral nerves. II Biochemical study. *Brain Res.* 1981; 219(1): 73-84.

Buttermore ED, Dupree JL, Cheng J, et al. The cytoskeletal adaptor protein band 4.1B is required for the maintenance of paranodal axoglial septate junctions in myelinated axons. *J. Neurosci.* 2011; 31(22): 8013-24.

Cerqueira JJ, Mailliet F, Almeida OF, et al. The prefrontal cortex as a key target of the maladaptive response to stress. *J. Neurosci.* 2007; 27(11): 2781-7.

Chiti Z, Teschemacher AG. Exocytosis of norepinephrine at axon varicosities and neuronal cell bodies in the rat brain. *FASEB J.* 2007; 21(10): 2540-50.

Choi SK, Lee GJ, Choi S, et al. Neuroprotective effects by nimodipine treatment in the experimental global ischemic rat model: real time estimation of glutamate. *J. Korean Neurosurg. Soc.* 2011; 49(1): 61-7.

Choo AM, Liu J, Liu Z, et al. Modeling spinal cord contusion, dislocation, and distraction: characterization of vertebral clamps, injury severities, and node of Ranvier deformations. *Neurosci. Methods* 2009; 181(1): 6-17.

Chowdhury SI, Coats J, Neis RA, et al. A bovine herpes virus type 1 mutant virus with truncated glycoprotein E cytoplasmic tail has defective anterograde neuronal transport in rabbit dorsal root ganglia primary neuronal cultures in a microfluidic chamber system. *J. Neurovirol.* 2010; 16(6): 457-65.

Cifuentes-Diaz C. Neurofilament accumulation at the motor endplate and lack of axonal sprouting in a spinal muscular atrophy mouse model. *Hum. Mol. Genet* 2002; 11(12): 1439-47.

Cook CS, Wellman CL. Chronic stress dendritic morphology rat medial prefrontal cortex. *J. Neurobiol.* 2004; 60(2): 236-48.

Cowdry EV. Laboratory Technique in biology and medicine. Baltimore: The Williams & Wilkins Company 1952: pp. 413.

Craddock TJ, Tuszynski JA, Priel A, et al. Microtubule ionic conduction and its implications for higher cognitive functions. *J. Integr. Neurosci.* 2010; 9(2): 103-22.

David AT, Saied A, Charles A, et al. A herpes simplex virus 1 (McKrae) mutant lacking the glycoprotein K gene is unable to infect via neuronal axons and egress from neuronal cell bodies. *MBio* 2012; 24(4): 144-212.

de Forges H, Bouissou A, Perez F. Interplay between microtubule dynamics and intracellular organization. *J. Biochem. Cell Biol.* 2012; 44(2): 266-74.

Deiters O. Untersuchungen über Gehirn und Rückenmark des Menschen und der Säugethiere. Edited by Max Schultze. Braunschweig: Vieweg und Sohn 1865: S. 318.

Deiters VS, Guillery RW. Otto Friedrich Karl Deiters (1834-1863). *J. Comp. Neurol.* 2013; 521(9): 1929-53.

Demoor J. La plasticite morphologique des neurones cerebraux. *Arch. Biol.* 1896; 14: 723-52.

De Regge N, Nauwynck HJ, Geenen K, et al. Alpha-herpes virus glycoprotein D interaction with sensory neurons triggers formation of varicosities that serve as virus exit sites. *J. Cell Biol.* 2006; 174(2): 267-75.

de Renyi GSt. The structure of cells in tissue as revealed by microdissection. II. The physical properties of the living axis cylinder in the myelinated nerve fiber of the frog. *J. Comp. Neurol.* 1929a; 47: 405-25.

de Renyi GSt. The structure of cells in tissue as revealed by microdissection. IV. Observations on neurofibrils in the living nervous tissue of the lobster (Homarus americanus). *J. Comp. Neurol.* 1929b; 48: 441-57.

Dewan AK, Ramey ML, Tricas TC. Arginine vasotocin neuronal phenotypes, telencephalic fiber varicosities, and social behavior in butterflyfish (Chaetodontidae): potential similarities to birds and mammals. *Horm. Behav.* 2011; 59(1): 56-66.

Dewan AK, Tricas TC. Arginine vasotocin neuronal phenotypes and their relationship to aggressive behavior in the territorial monogamous multiband butterflyfish, Chaetodon multicinctus. *Brain Res.* 2011; 1401: 74-84.

Dispersyn G, Nuydens R, Borgers M, et al. Nimodipine and flunarizine have different effects on survival and morphology of PC12 cells during nerve growth factor deprivation. *Europ. J. Pharmacol.* 1999; 384(1): 61-70.

Dogel AS. The Elements of the Nervous System and a New Theory of Sleep. *Natural Geography* 1896; 1(3): 229-54.

Dolbeare D, Houle JD. Restriction of axonal retraction and promotion of axonal regeneration by chronically injured neurons after intraspinal treatment with glial cell line-derived neurotrophic factor (GDNF). *J. Neurotrauma* 2003; 20(11): 1251-61.

Dolgo-Saburov BA. The Innervation of the Veins. Leningrad: Medgiz 1958; pp. 268.

Douvaras P, Wang J, Zimmer M, et al. Efficient generation of myelinating oligodendrocytes from primary progressive multiple sclerosis patients by induced pluripotent stem cells. *Stem Cell Reports* 2014; 3(2): 250-9.

Duval MR. Theorie mecaniquede la paralysie hysterifue, de somnambulisme du sommeil naturel et de la distraction. Comptes Rendus Heldomadaires. *Seances et memoires* 1895: 85-6.

Eccles JC. Discussion. Axoplasmic streaming in regenerating and in normal nerve fibres. *Prog. Brain Res.* 1964; 13: 1-71.

Eiland L, Ramroop J, Hill MN, et al. Chronic juvenile stress produces corticolimbic dendritic architectural remodeling and modulates emotional behavior in male and female rats. *Psychoneuroendocrinology* 2012; 37(1): 39-47.

El Maarouf A, Kolesnikov Y, Pasternak G, et al. Polysialic acid-induced plasticity reduces neuropathic insult to the central nervous system. *Proc. Natl. Acad. Sci. USA* 2005; 102(32): 11516-20.

Erfanian R, Firouzi M, Nabian MH, et al. Comparison of a new single-donor human fibrin adhesive with suture for posterior tibial nerve repair in rat: biomechanical resistance and functional analysis. *Chin J. Traumatol.* 2014; 1(3): 146-152.

Falnikar A, Tole S, Baas PW. Kinesin-5, a mitotic microtubule-associated motor protein, modulates neuronal migration. *Mol. Biol. Cell* 2011; 22(9): 1561-74.

Feller KD, Cronin TW. Hiding opaque eyes in transparent organisms: a potential role for larval eyeshine in stomatopod crustaceans. *J. Exp. Biol.* 2014; 217(Pt 18): 3263-73.

Fenoglio KA, Brunson KL, Baram TZ. Hippocampal neuroplasticity induced by early-life stress: functional and molecular aspects. *Front Neuroendocrinol.* 2006; 27(2): 180-92.

Fernandez-Moran H. The submicroscopic organisation of vertebrate nerve fibers. *Exp. Cell Res.* 1952; 3: 282-359.

Fernandets-Moran H. The Fine Structure of the Plate Systems of Biological Origin. In the book: *Modern Problems of Biophysics* 1961: pp. 58-68.

Fernandets- Moran H, Finean JB. Study of the Myelin Sheath by Electron Microscopy and Small Angle X-ray Analysis. In the book: *Questions of the Electron Microscopy of Tissue* 1959: pp. 67-96.

Filatov LA. About Fedor the Archer, a Swashbuckling Young Man. In the book: The Love for Three Oranges. Moscow: Eksmo-Press 1998: pp 11-89.

Fraczek-Szczypta A. Carbon nanomaterials for nerve tissue stimulation and regeneration. Mater Sci Eng 2014; 1(34): 35-49 Friedrichsberg DA. A *Course in Colloidal Chemistry. Leningrad: Chemistry* 1984: pp. 368.

Freedman H, Rezania V, Priel A, et al. Model of ionic currents through microtubule nanopores and the lumen. *Phys. Rev. E Stat Nonlin Soft Matter Phys.* 2010; 81(5 Pt 1): 051912.

Frolov Y. A Course in Colloidal Chemistry. *Surface Phenomena and Dispersal Systems. Moscow: Chemistry* 1982: pp. 400.

Gardiner J, Overall R, Marc J. The microtubule cytoskeleton acts as a key downstream effector of neurotransmitter signaling. *Synapse* 2011; 65(3): 249-56.

Gaudet AD, Popovich PG, Ramer MS. Wallerian degeneration: gaining perspective on inflammatory events after peripheral nerve injury. *J. Neuroinflammation* 2011; 8: 110.

Gencer ZK, Özkiriş M, Saydam L, et al. The comparison of histological results of experimentally created facial nerve defects repaired by 2 different anastomosis techniques: classic suture technique or tissue adhesives for nerve anastomosis? *J. Craniofac. Surg.* 2014; 25(2): 652-6.

Giachello CN, Montarolo PG, Ghirardi M. Synaptic functions of invertebrate varicosities: what molecular mechanisms lie beneath. *Neural. Plast.* 2012: 1-14.

Golgi C. Sulle alterazioni degli centrali nervousi in un caso di corea gesticolatoria associates ad alienazione mentale. *Rev. Clin. Bologia* 1874; 2: 361-77.

Greenberg ML, Weinger JG, Matheu MP, et al. Two-photon imaging of remyelination of spinal cord axons by engrafted neural precursor cells in a viral model of multiple sclerosis. *Proc. Natl. Acad. Sci. USA* 2014; 111(22): 2349-55.

Grutzendler J, Kasthuri N, Gan WB. Long-term dendritic spine stability in the adult cortex. *Nature* 2002; 420(6917): 812-6.

Gulyaev NV. The Impact of Stress Factors on the Functioning of the Hippocampus of the Adult Organism: Molecular and Cellular Mechanisms of the Dorsoventral gradient. *Russian Physiological Magazine* 2013; 99(1): 3-16.

Hamon L, Savarin P, Curmi PA, et al. Rapid assembly and collective behavior of microtubule bundles in the presence of polyamines. *Biophys* 2011; 101(1): 205-16.

Hanlon R. Cephalopod dynamic camouflage. *Curr. Biol.* 2007; 17(11): 400-4.

Hao J, Rapp PR, Leffler AE, et al. Estrogen alters spine number and morphology in prefrontal cortex of aged female rhesus monkeys. *J. Neurosci.* 2006; 26(9): 2571-8.

Hasaka TP, Myers KA, Baas PW. Role of actin filaments in the axonal transport of microtubules. *J. Neurosci.* 2004; 24(50): 11291-301.

Hellal F, Hurtado A, Ruschel J, et al. Microtubule stabilization reduces scarring and causes axon regeneration after spinal cord injury. *Science* 2011; 331(6019): 928-31.

Hill MH, Hillard CJ, McEwen BS. Alterations in corticolimbic dendritic morphology and emotional behavior in cannabinoid CB1 receptor-deficient mice parallel the affects of chronic stress. *Cereb. Cortex* 2011; 21(9): 2056-64.

Hirokawa N, Shiomura Y, Okabe S. Tau proteins: the molecular structure and mode of binding on microtubules. *J. Cell Biol.* 1988; 107(4): 1449-59.

Hollenbeck PJ. Directing traffic and autophagy in axonal transport. *Dev. Cell* 2014; 29(5): 505-6.

Huh JW, Laurer HL, Raghupathi R, et al. Rapid loss and partial recovery of neurofilament immunostaining following focal brain injury in mice. *Exp. Neurol.* 2002; 175(1): 198-208.

Izquierdo A, Wellman CL, Holmes A. Brief uncontrollable stress causes dendritic retraction in infralimbic cortex and resistance to fear extinction in mice. *J. Neurosci.* 2006; 26(21): 5733-8.

Jones M. Biochemical Thermodynamics. M.: *Mir* 1982; pp. 440.

Jung C, Chylinski TM, Pimenta A, et al. Neurofilament transport is dependent on actin and myosin. *J. Neurosci.* 2004; 24(43): 9486-96.

Kaivarainen AI. The Dynamic Behavior of Proteins in Aqueous Medium and their Functions. Leningrad: *Science* 1980; pp. 271.

Kalyunov VN. Biology of Growth Factor in Nerve Tissue. *Minsk, Science and Technology* 1986; pp. 197.

Karcer SJ. Molecular biologi. London: Academic Press 1995: pp. 106.

Kasai H, Matsuzaki M, Noguchi J, et al. Structure-stability-function relationships of dendritic spines. *Trends Neurosci.* 2003; 26(7): 360-8.

Kato S, Kobayashi K, Inoue K, et al. A lentiviral strategy for highly efficient retrograde gene transfer by pseudotyping with fusion envelope glycoprotein. *Hum. Gene Ther.* 2011; 22(2): 197-206.

Kilinc D, Gallo G, Barbee KA. Mechanical membrane injury induces axonal beading through localized activation of calpain. *Exp. Neurol.* 2009; 219(2): 553-61.

Kirchhoff GR. Untersuchungen über das Sonnenspectrum und die Spectren der chemischen Elemente. Berlin 1861: S. 576.

Kislov AN, Chekurova NR, Kataev AA. Osmotic Gradient and Modulators of the Cytoskeleton in the Functional Properties of Nicotinic Receptors of Neurons of the Pond Snail. *Biophysics of Cells and Cell to Cell Interaction: Moscow:* Moscow 1999; pp. 48.

Klimaschewski L, Hausott B, Angelov DN. The pros and cons of growth factors and cytokines in peripheral axon regeneration. *Int. Rev. Neurobiol.* 2013; 108: 137-71.

Knafo S, Libersat F, Barkai E. Olfactory learning-induced morphological modifications in single dendritic spines of young rats. *Eur J. Neurosci.* 2005; 21(8): 2217-26.

Kokurina TN, Sotnikov OS, Novakovskaia SA, et al. Interdependent changes in axon and Schwann cells during reactive rearrangements of myelinated fibers. *Neurosci. Bihavioral Physiology* 2014; 44(5): 528-35.

Kramer T, Greco TM, Taylor MP, et al. Kinesin-3 mediates axonal sorting and directional transport of alpha herpes virus particles in neurons. *Cell Host Microbe* 2012; 12(6): 806-14.

Kristenson K, Olsson J. Retrograde transport of protein. *Brain Res.* 1971; 29(2): 363-5.

Kuznetsov SA. Large Explanatory Dictionary. SPB: Norint 2002; pp. 1536.

Kuznetsov IA, Kuznetsov AV. Modeling anterograde and retrograde transport of short mobile microtubules from the site of axonal branch formation. *J. Biol. Phys.* 2014; 40(1): 41-53.

Lazek RJ. Bidirectinal transport of radio-actively labelled axoplasmic components. *Nature* 1961; 216(5121): 1212-4.

Lerch JK, Martínez-Ondaro YR, Bixby JL, et al. c Jun promotes CNS axon growth. *Mol. Cell Neurosci.* 2014; 59: 97-105.

Lesgaft PF. Selected works on Anatomy. Moscow 1968.

Lewin B, Kassimeris L, Lingappa VP, et al. *Cells.* Moscow: Binom 2011: pp. 651.

Leydig F. Lehrbuch der Histologia des Menschen und der Thiere 1857: pp. 51.

Li Y, Hu X, Liu Y, et al. Nimodipine protects dopaminergic neurons against inflammation-mediated degeneration through inhibition of microglial activation. *Neuropharmacology* 2009; 56(3): 580-9.

Liu XY, Liang H, Deng JY, et al. Analysis of clinicopathological characteristics and prognosis of patients with gastric cancer in different locations after radical resection. *Zhonghua Wei Chang Wai Ke Za Zhi* 2010; 13(3): 181-4.

Liu M, Nadar VC, Kozielski F, et al. Kinesin-12, a mitotic microtubule-associated motor protein, impacts axonal growth, navigation, and branching. *J. Neurosci.* 2010; 30(44): 14896-906.

Lowery LA, Vactor DV. The trip of the tip: understanding the growth cone machinery. *Nat. Rev. Mol. Cell Biol.* 2009; 10(5): 332-43.

Lubinska L. Axoplasmic streaming in regenerating and in normal nerve fibres. *Prog Brain Res* 1964; 13: 1-71.

Lubinska L, Lukaszewska I. Shape of myelinated nerve fibres and proximodistal flow of axoplasm. *Acta Biol. Exp.* (Warszawa) 1956; 17: 115-33.

Lubinska L, Niemierko S, Oberfeld B. Gradient of cholinesterase activity. *Nature* 1961; 189: 122-3.

Lubinska L, Niemierko S, Oberfeld B, et al. Bidirectional movements of axoplasm in peripheral nerve fibres. *Acta Biol. Exp.* (Warszawa) 1963; 23: 239-47.

Lukashin VG, Podolsky LA, Soloviev NA. Living Bush-like Receptors and an Analysis of their Morphological and Functional State. Mechanisms of the Neuron Response to Irritating Effects. Leningrad: *Science* 1981: 6-11.

Magarifios AM, Li CJ, Gal Toth J, et al. Effect of brain-derived neurotrophic factor haploinsufficiency on stress-induced remodeling of hippocampal neurons. *Hippocampus* 2011; 21(3): 253-64.

Malkinson G, Spira ME. Clustering of excess growth resources within leading growth cones underlies the recurrent "deposition" of varicosities along developing neurites. *Exp. Neurol.* 2010; 225(1): 140-53.

Mandelkow EM, Mandelkow E. Biochemistry and cell biology of tau protein in neurofibrillary degeneration. *Cold Spring Harb Perspect Med.* 2012; 2(7): 1-25.

Mann F, Rougon G. Mechanisms of axon guidance: membrane dynamics and axonal transport in semaphorin signalling. *J. Neurochem.* 2007; 102: 316-23.

Martínez-Tellez R, Gómez-Villalobos MJ, Flores G. Alteration in dendritic morphology of cortical neurons in rats with diabetes mellitus induced by streptozotocin. *Brain Res.* 2005; 1048(1-2): 108-15.

Martins RS, Teodoro WR, Simplicio H, et al. Influence of suture on peripheral nerve regeneration and collagen production at the site of neurorrhaphy: an experimental study. *Neurosurgery* 2011; 68(3): 765-72.

Masurovsky EB, Peterson ER, Crain SM, et al. Microtubule arrays in taxol-treated mouse dorsal root ganglion-spinal cord cultures. *Brain Res.* 1981; 217(2): 392-8.

Matveev AN. Electricity and Magnetism. Textbook. Moscow: High School 1983: pp. 463.

Mauthner L. Recherches sur la structure du systeme nerveux. Paris 1868.

Miller MM, Morrison JH, McEwen BS. Basal anxiety-like behavior predicts differences in dendritic morphology in the medial prefrontal cortex in two strains of rats. *Behav. Brain Res.* 2012; 229(1): 280-8.

Mudrakola HV, Zhang K. Optically resolving individual microtubules in live axons. *Structure* 2009; 17(11): 1433-41.

Mussel M, Zeevy K, Diamant H, et al. Drag of the cytosol as a transport mechanism in neurons. *Biophys. J.* 2014; 106(12): 2710-9.

Myers KA, Tint I, Nadar CV, et al. Antagonistic forces generated by cytoplasmic dynein and myosin-II during growth cone turning and axonal retraction. *Traffic* 2006; 7(10): 1333-51.

Nguyen MD, Shu T, Sanada K, et al. A NUDEL-dependent mechanism of neurofilament assembly regulates the integrity of CNS neurons. *Nat. Cell Biol.* 2004; 6(7): 595-608.

Nicholls JG, Martin AR, Wallace BD, et al. *From the Neuron to the Brain.* Moscow: LKI 2008: pp. 672.

Niculescu IT. Normal Nerve Fiber. *Pathology of the Nervous System* 1963: pp 980-91.

Nimchinsky EA, Sabatini BL, Svoboda K. Structure and function of dendritic spines. *Annu. Rev. Physiol.* 2002; 64: 313-53.

Papasozomenos SC, Payne MR. Actin immunoreactivity localizes with segregated microtubules and membraneous organelles and in the subaxolemmal region in the beta, beta'-iminodipropionitrile axon. *Neurosci.* 1986; 6(12): 3483-91.

Payne SC, Bartlett CA, Harvey AR, et al. Myelin sheath decompaction, axon swelling, and functional loss during chronic secondary degeneration in rat optic nerve. *Invest Ophthalmol. Vis. Sci.* 2012; 53(10): 6093-101.

Perrin-Tricaud C, Rutishauser U, Tricaud N. P120 catenin is required for thickening of Schwann cell myelin. *Mol. Cell Neurosci.* 2007; 35(1): 120-9.

Piao ZX, Wang WS, Xu XJ, et al. Autophagy of neuron axon during regeneration of rat sciatic nerves. *Di Yi Jun Yi Da Xue Xue Bao* 2004; 24(4): 361-4.

Plateau AF. Statique experimentale et theorique des liquids soumis aux seulis forces moleculaires. Gland; Leipzig: Clemm 1973; 1: pp. 450.

Popov VI, Bocharova LS. Hibernation-induced structural changes in synaptic contacts between mossy fibres and hippocampal pyramidal neurons. *Neuroscience* 1992; 48(1): 53-62.

Popov VI, Bocharova LS, Bragin AG. Repeated changes of dendritic morphology in the hippocampus of ground squirrels in the course of hibernation. *Neuroscience* 1992; 48(1): 45-51.

Pugh PC, Berg DK. Neuronal acetylcholine receptors that bind alpha-bungarotoxin mediate neurite retraction in a calcium-dependent manner. *J. Neurosci.* 1994; 14(2): 889-96.

Qiang L, Yu W, Liu M, et al. Basic fibroblast growth factor elicits formation of interstitial axonal branches via enhanced severing of microtubules. *Mol. Biol. Cell* 2010; 21(2): 334-44.

Radley JJ, Rocher AB, Rodriguez A, et al. Repeated stress alters dendritic spine morphology in the rat medial prefrontal cortex. *J. Comp. Neurol.* 2008; 507(1): 1141-50.

Radtke C, Allmeling C, Waldmann KH, et al. Spider silk constructs enhance axonal regeneration and remyelination in long nerve defects in sheep. *PLoS One* 2011; 6(2): 1-10.

Ramon y Cajal S. Degeneration, regeneration of the nervous system. N.Y.: Haften Publishing CO 1959: pp. 123.

Ramon y Cajal S. Histilogy of the nervous system of man and vertebrates. New York; Oxford: Oxford Univ. Press 1995; 2: pp. 805.

Rao MV, Garcia ML, Miyazaki Y, et al. Gene replacement in mice reveals that the heavily phosphorylated tail of neurofilament heavy subunit does not affect axonal caliber or the transit of cargoes in slow axonal transport. *J. Cell Biol.* 2002; 158(4): 681-93.

Ren J, Zhou X, Galligan JJ. 5-HT4 receptor activation facilitates recovery from synaptic rundown and increases transmitter release from single varicosities of myenteric neurons. *Am. J. Physiol. Gastrointest Liver Physiol.* 2008; 294(6): 1376-83.

Retzius G. Weitere Beiträge zur Kenntnis der Cajalschen Zellen der Grosshirnrinde des Menschen. *Biol. Untersuch* 1894; 6: 29-34.

Roth S, Bisbal M, Brocard J, et al. How morphological constraints affect axonal polarity in mouse neurons. *PLoS One* 2012; 7(3): 1-10.

Roy S. Seeing the unseen: the hidden world of slow axonal transport. *Neuroscientist* 2014; 20(1): 71-81.

Salinas S, Bilsland LG, Henaff D, et al. CAR-associated vesicular transport of an adenovirus in motor neuron axons. *PLoS Pathog.* 2009; 5(5): 1-16.

Sasaki-Sherrington SE, Jacobs JR, Stevens JK. Intracellular control of axial shape in non-uniform neurites: a serial electron microscopic analysis of organelles and microtubules in AI and AII retinal amacrine neurites. *J. Cell Biol.* 1984; 98(4): 1279-90.

Sauter E, Buckwalter JA, McKinley TO, et al. Cytoskeletal dissolution blocks oxidant release and cell death in injured cartilage. *J. Orthop. Res.* 2012; 30(4): 593-8.

Senda T, Fujita H. Ultrastructural aspects of quick-freezing beep-etching replica images of the cytoskeletal. System in anterior pituitary secretory cells of rats and mice. *Arch. Histol. Jap.* 1987; 50: 49-60.

Shapovalov VM, Dulaev AK, Shulev YA, et al. Orthopedic Aspects of Surgical Treatment of Patients with Degenerative-dystrophic Diseases of the Lumbosacral Area of the Spine . *Spinal Surgery* 2005; 3: 61-70.

Shen D, Wang X, Gu X. Scar-modulating treatments for central nervous system injury. *Neurosci. Bull* 2014: 24.

Skibo GG. Current concepts of the mechanisms of axonal transport. *Neirofiziologiia* 1980; 12(4): 432-43. (in Rus).

Sotnikov OS. Functional Morphology of Living Myelinated Nerve Fibers. Leningrad: *Science* 1976: pp. 199.

Sotnikov OS. Dynamics of the Structure of the Living Neuron. Leningrad: *Science* 1985: pp. 219.

Sotnikov OS. Statics and Dynamics of Asynaptic Dendrites. St. Petersburg: *Nauka* 2008: pp. 397.

Sotnikov OS. Use of cell culture to prove syncytial connection and fusion of neurons. Biomedical Tissue Culture Ch. 6. Ed by L. Ceccherini-Nelli, B Matteoli Croatia, Intech 2012: 83-114.

Sotnikov OS, Kokurina TN. Water translocation from the axoplasm into the glioplasm during reorganisation of myelinated nerve fibres. Axons: Cell Biology, Molecular Dynamics and Roles in Neural Repair and Rehabilitation. New York: “NOVA Biomedical” 2013; Ch. 2: 95.

Sotnikov OS, Malashko VV, Rybakov GI. The Phenomenon of Fusion of Nerve Fibers. *Reports of the Russian Academy of Sciences* 2006; 410(1): 130-3.

Sotnikov OS, Vasiagina NYu, Sergeeva SS. Traumatic retraction of living neural processes and its inhibition. Axons: Cell Biology, Molecular Dynamics and Roles in Neural Repair and Rehabilitation. New York: “NOVA Biomedical” 2013; Ch. 1: 1-94.

Sotnikov OS, Vasyagina NYu, Sergeeva SS. Simultaneous opposite axonal currents in neural process. *Biophysics* 2014a; 50(5): 746.

Sotnikov OS, Vasyagina NYu, Sergeeva SS, et al. Simultaneous axonal flows of opposite direction in neurites. Biological Motility. New facts and hypotheses. *Pushchino: Russian Academy of Sciences* 2014b: 289-92.

Soukhanoff SA. Contribution a l’etude des modifications que subissent les prolongements dendritiques des Cellules nerveuses sous l’influence des narcotique. *La Cellule* 1898a; 14: 388-95.

Soukhanoff SA. D’Anatomie pathologique de la Cellule nerveuse en rapport avec l’Atrophie variqueuse des Dentrites de l’Ecorce cerebrale. *La Cellule* 1898b; 14: 399-415.

Soukhanoff SA. Sur l'etat variqueux des dendrites corticales. *Arch Neurol* 1899; 9: 273-89.

Stefanowska M. Les appendices terminaux des dendrites cérébraux et leurs differents états physiologiques. *Travaux de laboratoire de l'Institut Solvay Bruxelles fascic* 1897; III: 18-29.

Stefanowska M. Evolution des cellules nerveuses cortic des chez la soures après la naissance. Traux de Laboratoire de L'Institut Solvey Bruxelles 1898; 1(II): fascicule 2.

Stehr F. Textbook of Histology. 3rd Russian edition. St. Petersburg: Publishing House of KL Ricker 1908: pp. 688.

Susuki K, Raphael AR, Ogawa Y, et al. Schwann cell spectrins modulate peripheral nerve myelination. *Proc. Natl. Acad. Sci. USA* 2011; 108(19): 8009-14.

Suter DM, Miller KE. The emerging role of forces in axonal elongation. *Prog. Neurobiol.* 2011; 94(2): 91-101.

Swetman CA, Leverrier Y, Garg R, et al. Extension, retraction and contraction in the formation of a dendritic cell dendrite: distinct roles for Rho GTPases. *Eur. J. Immunol.* 2002; 32(7): 2074-83.

Taylor D, Grimm H, Stout W. Biology. Moscow: Mir 2007; 1: pp. 194.

Terada N, Saitoh Y, Ohno N, et al. Essential function of protein 4.1G in targeting of membrane protein palmitoylated 6 into Schmidt-Lanterman incisures in myelinated nerves. *Mol. Cell Biol.* 2012; 32(1): 199-205.

Tischfield MA, Baris HN, Wu C; et al. Human TUBB3 mutations perturb microtubule dynamics, kinesin interactions, and axon guidance. *Cell* 2010; 140(1): 74-87.

Totibadze NK. The Work of the Institute of Physiology. *ANT SSR* 1958; 11: 231-4.

Turgut M, Uysal A, Pehlivan M, et al. Assessment of effects of pinealectomy and exogenous melatonin administration on rat sciatic nerve suture repair: an electrophysiological, electron microscopic, and immunohistochemical study. *Acta Neurochir.* (Wien) 2005; 147(1): 67-77.

Velázquez-Bernardino P, García-Sierra F, Hernández-Hernández O, et al. Myotonic dystrophy type 1-associated CTG repeats disturb the expression and subcellular distribution of microtubule-associated proteins MAP1A, MAP2, and MAP6/STOP in PC12 cells. *Mol. Biol. Rep* 2012; 39(1): 415-24.

von der Ohe CG, Smith CD, Garner CC, et al. Ubiquitous and temperature – dependent neural plasticity in hibernators. *J. Neurosci.* 2006; 26(41): 10590-8.

Voronin VV, Kuparadze MR, Totibadze NK. Myelinated Nerve Fibers. Tbilisi: Metsniereba 1960; pp. 53.

Wang Q, Chen L, Chen L, et al. The tau-like protein in silkworm (Bombyx mori) induces microtubule bundle formation. *Front Biosci.* (Elite Ed) 2012; 4: 998-1008.

Weiss PA. Parameters of nerve regeneration. In the book: The Regeneration of the Central Nervous System, Moscow 1959; pp. 100-3.

Weiss PA. The cell in development. *Lab. Invest* 1959; 8(2): 415-30.

Weiss PA. Neuronal dynamics and neuroplasmic flow. The Neurosciences. N. Y.: Second study program 1970: 840-50.

Weiss PA. Neuronal dynamics and axonal flow: axonal peristalsis. *Proc. Nat. Acad. Sci. USA* 1972; 69(5): 1309-12.

Weiss PA, Hiscoe HB. Experiments on the mechanism of nerve growth. *J. Exp. Zool.* 1948; 107: 315-95.

Weiss PA, Taylor AC, Pillai PA. The nerve fiber as a system in continuous flow: microcinematographic and electronmicroscopic demonstrations. *Science* 1962; 136(3513): 330.

Wells HG. The Invisible man. M.: Menedzer 1898: pp. 78.

White LA, Baas PW, Heidemann SR. Microtubule stability in severed axons. *J. Neurocytol.* 1987; 16(6): 775-84.

Wortman JC, Shrestha UM, Barry DM, et al. Axonal transport: how high microtubule density can compensate for boundary effects in small-caliber axons. *Biophys. J.* 2014; 106(4): 813-23.

Wylie SR, Chantler PD. Myosin IIA drives neurite retraction. *Mol. Biol. Cell* 2003; 14(11): 4654-66.

Yamasaki R, Lu H, Butovsky O, et al. Differential roles of microglia and monocytes in the inflamed central nervous system. *J. Exp. Med.* 2014; 211(8): 1533-49.

Zaichick SV, Bohannon KP, Hughes A, et al. The herpes virus VP1/2 protein is an effector of dynein-mediated capsid transport and neuroinvasion. *Cell Host Microbe* 2013; 13(2): 193-203.

Zenbush P. Molecular Cell Biology. Volume 3, Moscow: Mir 1982; pp. 344.

Zhabotinsky YM. Normal and Pathological Morphology of the Neuron. Leningrad: Medicine 1965; pp. 323.

Zhang ZW, Kang JI, Vaucher E. Axonal varicosity density as an index of local neuronal interactions. *PLoS One* 2011; 6(7): 1-11.

Zylinski S, How MJ, Osorio D, et al. To be seen or to hide: visual characteristics of body patterns for camouflage and communication in the Australian giant cuttlefish Sepia apama. *Am. Nat.* 2011; 177(5): 681-90.

AUTHOR CONTACT INFORMATION

O. S. Sotnikov
Professor, Dr. Biology
Head of the Laboratory
Laboratory of Functional Morphology
and Physiology of the Neuron
Pavlov Institute of Physiology of the Russia Academy
of Sciences, Russia
E-mail: ossotnikov@mail.ru

INDEX

A

B

C

D

E

F

G

H

I

K

L

M

N

O

P

U

V

W